Solar Photovoltaic

RENEWABLE ENERGIES SERIES

skills2learn
www.skills2learn.com
Experts in e-learning & virtual reality simulation

CENGAGE
Learning·

Australia • Brazil • Japan • Korea • Mexico • Singapore • Spain • United Kingdom • United States

Solar Photovoltaic, 1st Edition
Skills2Learn

Publishing Director: Linden Harris

Commissioning Editor: Lucy Mills

Development Editor: Lauren Darby

Senior Project Editor: Alison Burt

Senior Manufacturing Buyer: Eyvett Davis

Typesetter: MPS Limited

Cover design: HCT Creative

For product information and technology assistance,
contact **emea.info@cengage.com**.

For permission to use material from this text or product,
and for permission queries,
email **emea.permissions@cengage.com**.

British Library Cataloguing-in-Publication Data

A catalogue record for this book is available from the British Library.

ISBN: 978-1-4080-6467-2

Cengage Learning EMEA

Cheriton House, North Way, Andover, Hampshire, SP10 5BE
United Kingdom

Cengage Learning products are represented in Canada by Nelson Education Ltd.

For your lifelong learning solutions, visit **www.cengage.co.uk**

Purchase your next print book, e-book or e-chapter at **www.cengagebrain.com**

Printed in China by RR Donnelley
1 2 3 4 5 6 7 8 9 10 – 15 14 13

Contents

Foreword

The energy sector is a significant part of the UK economy and a major employer of people. It has a huge impact on the environment and plays a massive role in our everyday life, shaping both our work and domestic habits and processes. With environmental issues such as climate change and sustainable sourcing of energy now playing an important role in our society, there is a need to educate a significant pool of people about the future technologies with renewable energies in all likelihood playing an increasingly significant part in our total energy requirements.

This environmental and renewable energy series of e-learning programmes and text workbooks has been developed to provide a structured blended learning approach that will enhance the learning experience and stimulate a deeper understanding of the renewable energy trades and give an awareness of sustainability issues. The content within these learning materials has been aligned as far as is currently possible to the units of the National Occupational Standards and can be used as a support tool whilst studying for any relevant vocational qualifications.

The uniqueness of this renewable energy series is that it aims to bridge the gap between classroom-based and practical-based learning. The workbooks provide classroom-based activities that can involve learners in discussions and research tasks as well as providing them with understanding and knowledge of the subject. The e-learning programmes take the subject further, with high quality images, animations and audio further enhancing the content and showing information in a different light. In addition, the e-practical side of the e-learning places the learner in a virtual environment where they can move around freely, interact with objects and use the knowledge and skills they have gained from the workbook and e-learning to complete a set of tasks whilst in the comfort of a safe working environment.

The workbooks and e-learning programmes are designed to help learners continuously improve their skills and provide confidence and a sound knowledge base before getting their hands dirty in the real world.

About e-Consortia

This series of renewable energies workbooks and e-learning programmes has been developed by the E-Renewable Consortium. The consortium is a group of colleges and organizations that are passionate about the renewable energy industry and are determined to enhance the learning experiences of people within the different trades or those that are new to it.

The consortium members have many years experience in the renewable energy and educational sectors and have created this blended learning approach of interactive e-learning programmes and text workbooks to achieve the aim of:

- Providing accessible training in different areas of renewable energy
- Bridging the gap between classroom-based and practical-based learning
- Providing a concentrated set of improvement learning modules
- Enabling learners to gain new skills and qualifications more effectively
- Improving functional skills and awareness of sustainability issues within the industry
- Promoting health and safety in the industry
- Encouraging training and continuous professional development.

For more information about this renewable energy consortium please visit: **http://skills2learn.cengage.co.uk/9-renewable-energy**

About e-learning

INTRODUCTION

This renewable energies series of workbooks and e-learning programmes uses a blended learning approach to train learners about renewable energy skills. Blended learning allows training to be delivered through different mediums such as books, e-learning (computer-based training), practical workshops, and traditional classroom techniques. These training methods are designed to complement each other and work in tandem to achieve overall learning objectives and outcomes.

E-LEARNING

The Solar PV e-learning programme that is also available to sit alongside this workbook offers a different method of learning. With technology playing an increasingly important part of everyday life, e-learning uses visually rich 2D and 3D graphics/animation, audio, video, text and interactive quizzes, to allow you to engage with the content and learn at your own pace and in your own time.

E-ASSESSMENT

Part of the e-learning programme is an e-assessment 'End test'. This facility allows you to be self-tested using interactive multimedia by answering questions on the e-learning modules you will have covered in the programme. The e-assessment provides feedback on both correctly and incorrectly answered questions. If answers are incorrect the learner is advised to revisit the learning materials they need to study further.

E-PRACTICAL

Part of the e-learning programme is an e-practical interactive scenario. This facility allows you to be immersed in a virtual reality situation where the choices you make affect the outcome. Using 3D technology, you can move freely around the environment, interact with objects, carry out tests and make decisions and mistakes until you have mastered the subject. By practising in a virtual environment you will not only be able to see what you've learnt but also analyze your approach and thought process to the problem.

BENEFITS OF E-LEARNING

Diversity – E-learning can be used for almost anything. With the correct approach any subject can be brought to life to provide an interactive training experience.

Technology – Advancements in computer technology now allow a wide range of spectacular and engaging e-learning to be delivered to a wider population.

Captivate and Motivate – Hold the learner's attention for longer with the use of high quality graphics, animation, sound and interactivity.

Safe Environment – E-Practical scenarios can create environments which simulate potentially harmful real-life situations or replicate a piece of dangerous equipment, therefore allowing the learner to train and gain experience and knowledge in a completely safe environment.

Instant Feedback – Learners can undertake training assessments which feedback results instantly. This can provide information on where they need to re-study or congratulate them on passing the assessment. Results and Certificates could also be printed for future records.

On-Demand – Can be accessed 24 hours a day, 7 days a week, 365 days of the year. You can access the content at any time and view it at your own pace.

Portable Solutions – Can be delivered via a CD, website or LMS. Learners no longer need to travel to all lectures, conferences, meetings or training days. This saves many man-hours in reduced travelling, cost of hotels and expenses amongst other things.

Reduction of Costs – Can be used to teach best practice processes on jobs which use large quantities of expensive materials. Learners can practise their techniques and boost their confidence to a high enough standard before being allowed near real materials.

SOLAR PV E-LEARNING

The aim of the solar PV e-learning programme is to enhance a learner's knowledge and understanding of solar photovoltaic installation and systems. The course content is aligned to units from the Environmental National Occupational Standards (NOS) so can be used for study towards certification.

The programme gives the learners an understanding of the different types of solar PV panels, as well as looking at sustainability, health and safety and functional skills in an interactive and visually engaging manner. It also provides a 'real-life' scenario where the learner can apply the knowledge gained from the tutorials in a safe yet practical way.

By using and completing this programme, it is expected that learners will:

- Describe the principles of a solar PV system and its use as a renewable energy source
- Explain how a PV system generates electricity
- Describe the types of PV systems currently available and how they can be fixed to or into buildings
- List the components of a solar PV system
- List the principles of solar PV system design
- Explain the preparatory work required for installation of a solar PV system
- Describe the installation process and requirements for a solar PV system
- Describe the opportunity to feed excess generated electricity back into the national grid
- List the requirements and know how to test, commission, handover and maintain a solar PV installation.

The e-learning programme is divided into the following learning modules:

- Getting Started
- Solar PV Overview

- Health and Safety
- Solar PV Basics
- Types of Solar PV Panels
- System Components
- System Design
- Pre-installation
- Installation
- PV Protection
- Commissioning
- Handover and Maintenance
- Interactive E-Practical Scenarios

THE RENEWABLE ENERGIES SERIES

As part of the renewable energies series the following e-learning programmes and workbooks are available. For more information please visit: **http://skills2learn.cengage.co.uk/9-renewable-energy**

- Introduction to Renewable Energies
- Heat Pumps
- Solar Thermal Hot Water
- Building Heat Loss Calculator (programme only)
- Solar Radiation Calculator (programme only)

About the NOS

The National Occupational Standards (NOS) provide a framework of information that outline the skills, knowledge and understanding required to carry out work-based activities within a given vocation. Each standard is divided into units that cover specific activities of that occupation. Employers, employees, teachers and learners can use these standards as an information, support and reference resource that will enable them to understand the skills and criteria required for good practice in the workplace.

The standards are used as a basis to develop many vocational qualifications in the United Kingdom for a wide range of occupations. This workbook and associated e-learning programme are aligned to the Environmental National Occupational Standards (NOS) by being designed against the Qualification and Credit Framework (QCF) units, which are developed from the NOS. Such a process is a requirement of the minimum technical competency (MTC) document for solar PV installers. Therefore this book and its associated e-learning support full training towards a certificate as recognized by all bodies offering routes to the Microgeneration Certification Scheme (MCS) as evidence of suitable training. The information within relates to the following units:

- Know the health and safety risks and safe systems of work associated with solar photovoltaic system installation work
- Know the requirements of relevant regulations/standards relating to practical installation, testing and commissioning activities for solar photovoltaic system installation work
- Know the fundamental differences between AC and DC circuits within solar photovoltaic systems
- Know the purpose of solar photovoltaic system components
- Know the types, silicon characteristics and typical conversion efficiencies of solar photovoltaic modules
- Know the fundamental design principles used to determine solar photovoltaic system module array size and position requirements

- Know the preparatory work and be able to undertake the preparatory work required for solar photovoltaic system installation work
- Know the layouts and the requirements for installing solar photovoltaic module arrays
- Know solar photovoltaic system DC and AC circuit installation layouts within the scope of the relevant engineering recommendation for grid tied systems
- Know solar photovoltaic system protection techniques and components
- Be able to install solar photovoltaic system components
- Know how to and be able to test and commission solar photovoltaic systems installations
- Know how to and be able to handover a solar photovoltaic systems installation.

SUMMARY OF THE ABOVE

Or in simplified English, this book and the e-learning training materials have been designed around the latest guidance from all the relevant bodies that will support a full solar PV training course and assessment process.

They share the knowledge in a simple-to-digest format that is more enjoyable to use and so more likely to be successful in sharing the important information required to install, commission, handover and maintain solar PV systems.

About the book

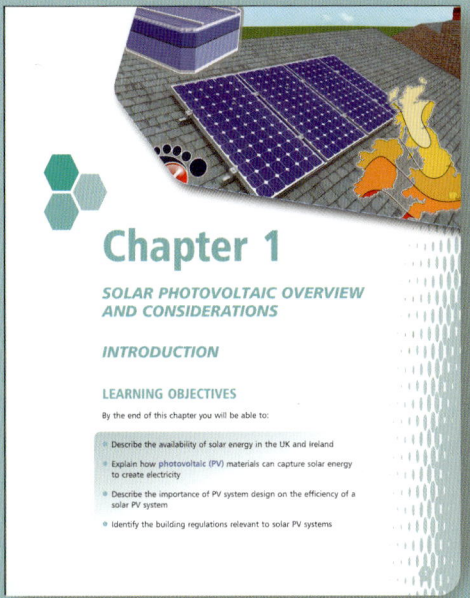

Learning Objectives at the start of each chapter explain the skills and knowledge you need to be proficient in and understand by the end of the chapter.

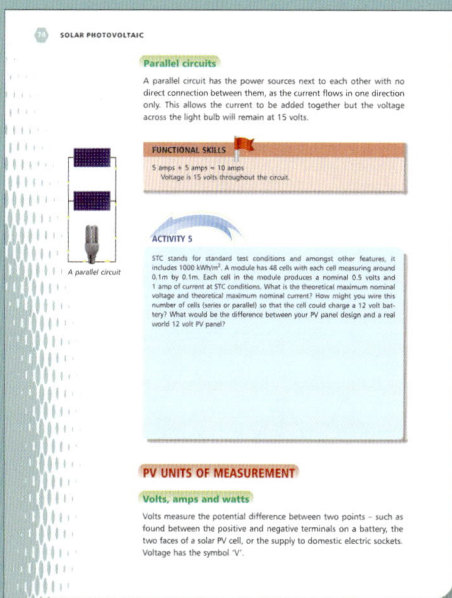

Activities are practical tasks that engage you in the subject and further your understanding.

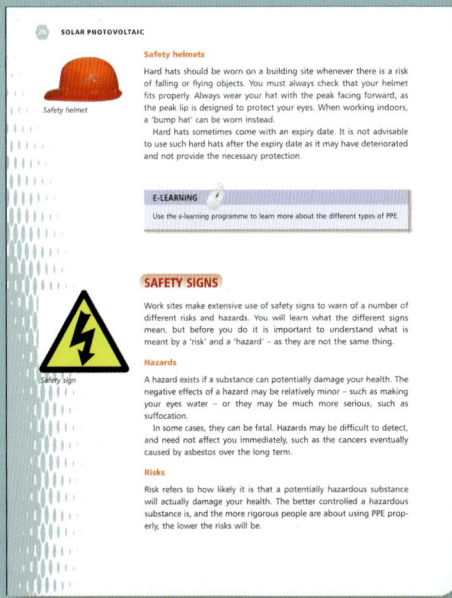

E-Learning Icons link the workbook content to the e-learning programme.

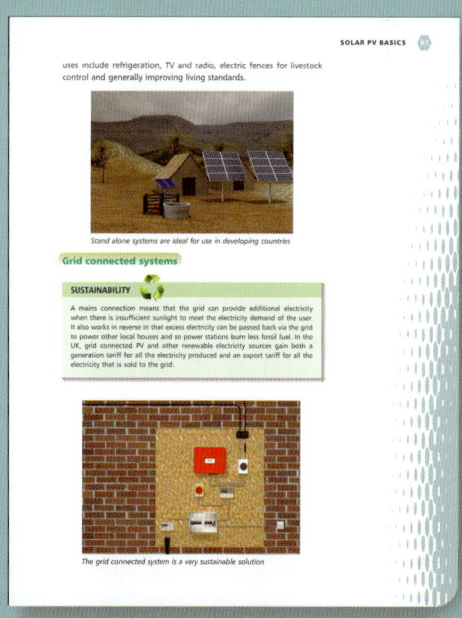

Sustainability Boxes provide information and helpful advice on how to work in a sustainable and environmentally friendly way.

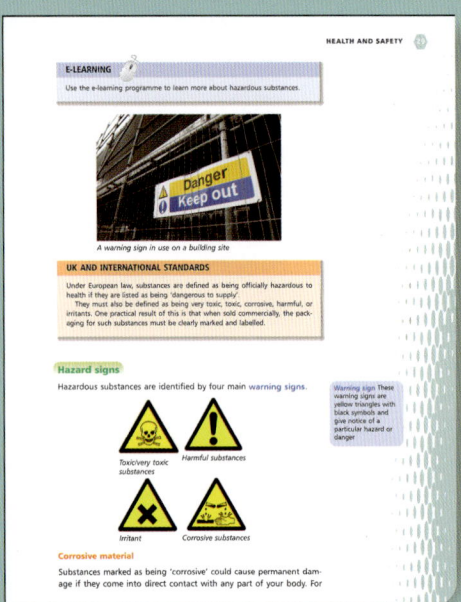

Note on UK Standards draws your attention to relevant building regulations.

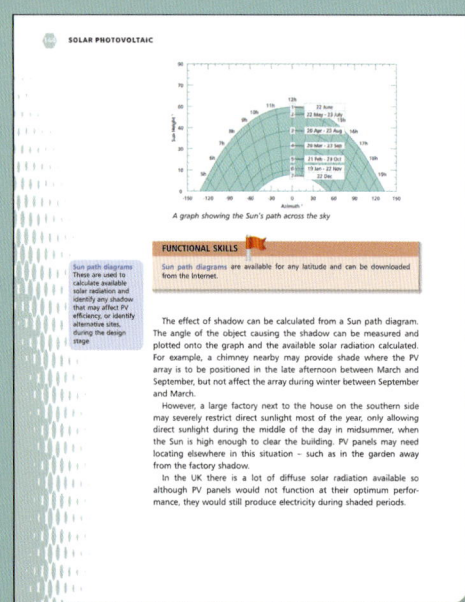

Functional Skills Icons highlight activities that develop and test your Maths, English and ICT key skills.

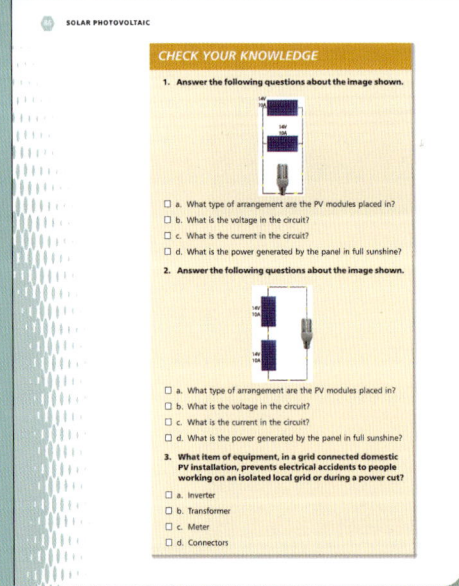

Check Your Knowledge at the end of each chapter to test your knowledge and understanding.

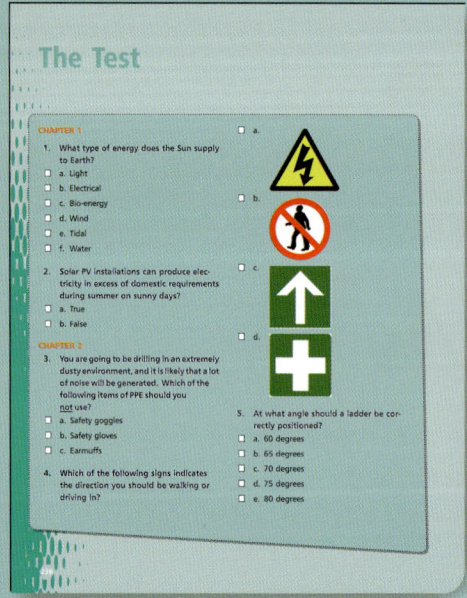

End Test in Chapter 12 checks your knowledge on all the information within the workbook.

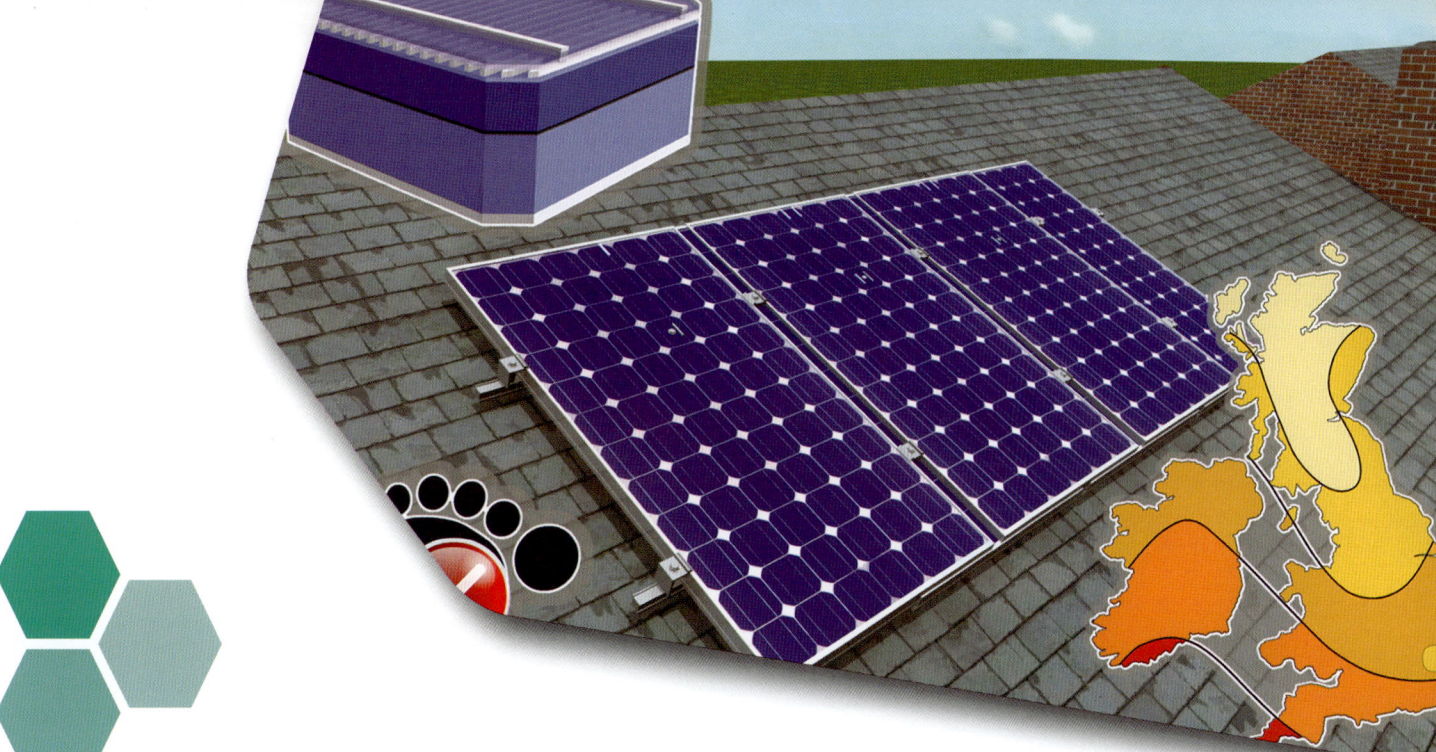

Chapter 1

SOLAR PHOTOVOLTAIC OVERVIEW AND CONSIDERATIONS

INTRODUCTION

LEARNING OBJECTIVES

By the end of this chapter you will be able to:

- Describe the availability of solar energy in the UK and Ireland

- Explain how **photovoltaic (PV)** materials can capture solar energy to create electricity

- Describe the importance of PV system design on the efficiency of a solar PV system

- Identify the building regulations relevant to solar PV systems

Satellite illustration of earth

PV or Photovoltaic
The process by which a device converts photons of solar energy directly into electricity

Solar energy The available solar energy is calculated as $1000\,W/m^2$ at the equator at midday

Solar radiation, solar irradiance Solar radiation at a location measured in $kWh/m^2/$annum. Irradiance is instant solar power in W/m^2. Solar radiation is also sometimes called insolation

SOLAR ENERGY IN THE UK

What is solar energy?

Solar energy is energy from the Sun and is a source of renewable energy, which is being developed as an alternative to fossil fuels such as coal, gas and oil. **Solar radiation** is regarded as an inexhaustible source of energy.

Solar panels

The Sun

The Sun is an inexhaustible source of energy. Radiant solar light energy from the Sun can be converted to both heat and electrical energy. Solar thermal energy can be used to heat water and buildings, and light energy can be used to produce electricity using photovoltaic panels.

Solar photovoltaic panels

Indirectly, solar energy also provides us with hydroelectric power, bio-energy, such as wood and other bio-fuels, wind power and wave power.

Wave energy

Hydro energy

Wind energy

Biomass energy

The Earth

There are two other types of renewable energy that are not derived from the Sun; geothermal energy from the Earth's core and tidal energy from the gravitational pull of the Moon and Sun.

Geothermal energy is derived from the Earth's core

Tidal energy is gravitational pull from the Moon and Sun

E-LEARNING

Use the e-learning programme to learn more about solar energy.

Solar radiation in the UK and Ireland

Different parts of the UK and Ireland receive different levels of solar radiation; the energy available decreases as you travel north. This variation is not only affected by latitude, but also the position of hills, mountains and trees. On top of these geographical variables there are also seasonal differences and annual variations created by the weather and how much sun is available, which can cause up to 10 per cent annual variation in system output.

All these factors affect the amount of energy that can be generated per m² of solar panel. Solar energy comes in two forms; the technical terms for these two forms are direct and diffuse.

Direct radiation

Direct radiation is sunshine, which is strongest in the summer at mid-day because the angle of the Sun is at its highest point. There is therefore a daily variation in direct radiation as well as a seasonal variation between summer and winter.

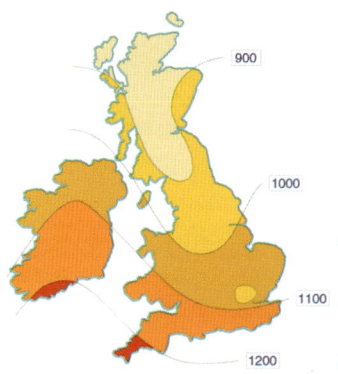

Figures shown on the map are average kWh/m² per year

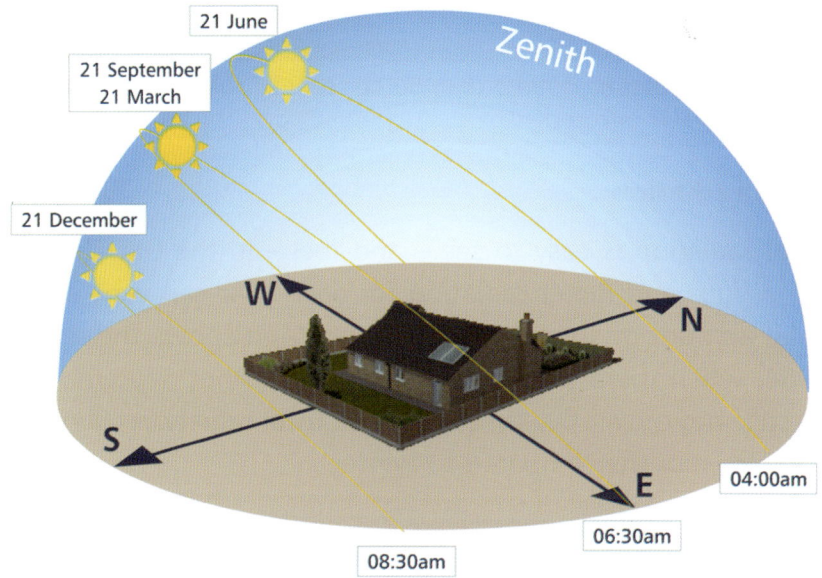

Illustration of direct radiation from the sun throughout the year

Diffuse radiation

Diffuse solar radiation gives us daylight and can be quite strong. When the Sun's rays hit the Earth's atmosphere some solar radiation is scattered in all directions and this is called diffuse radiation. It is assumed to come from all over the sky equally, rather than just the direction of the Sun. As the UK has many overcast days more than 50 per cent of solar radiation reaching a horizontal surface at ground level is diffuse. This type of solar radiation can still be used to power PV modules.

Diffuse solar radiation Diffuse solar radiation gives us daylight and can be quite strong. When the Sun's rays hit the Earth's atmosphere some radiation is scattered in all directions, creating diffuse radiation

Module A group of solar cells encapsulated in weatherproof glazing, sometimes with a junction box, cables and connectors. Usually factory fitted. Term used interchangeably with 'Panel'

Even on overcast days more than 50 per cent of solar radiation reaching ground level is diffused

Seasonal radiation

Seasonal radiation is easy to see – and feel – in the UK. The proportion of diffuse and direct radiation for each year can vary, as well as the total amount of solar radiation. This graph shows the recordings for 1964, when there was a poor summer and hot September, however it still shows over five times as much total radiation falling during the summer months compared to winter.

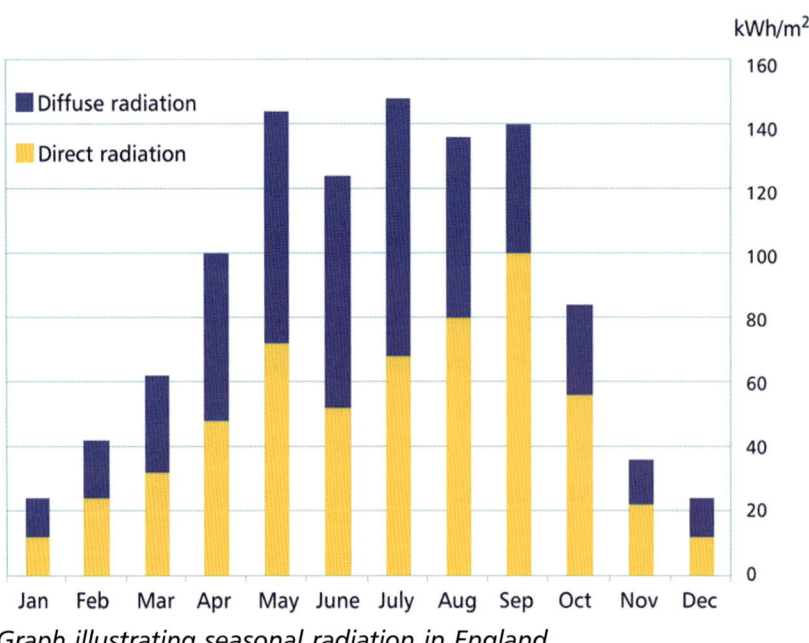

Graph illustrating seasonal radiation in England

E-LEARNING

Use the e-learning programme to learn more about the different types of radiation.

HOW SOLAR PV WORKS

An overview of photovoltaic cells

The word photovoltaic comes from photo which means light and voltaic which means electricity. Photovoltaic or PV **cells** convert the energy from the Sun directly into electricity. A group of cells connected together is called a module and a module packaged into a frame is more commonly known as a solar or PV panel. A group of PV panels is known as an **array.**

In the UK and Ireland we have a useful source of solar energy with around half of this energy being diffuse or indirect sunlight. Electricity can still be produced on cloudy or overcast days but the highest output will always come from the strongest, unobscured sunlight.

> **Cell** A PV cell is the smallest element within a PV module to convert light energy into DC electrical energy

PV cell

E-LEARNING

Use the e-learning programme to see a demonstration of PV cells.

> **Array** A number of connected PV panels functioning as a single electricity producing unit. The solar panels may be installed on a roof, wall, pole, or ground mounted frame

PV cells have many uses. They can be used in situations where there is no access to, or it is impractical to use, mains electricity. They are used to power appliances from satellites to watches. They are frequently used in the developing world to power water pumps and medical refrigerators.

UK AND INTERNATIONAL STANDARDS

More recently, the UK government is supporting use of solar PV and the domestic British market is rapidly growing. Similarly across most of Europe, as a result of the Renewable Energy Services directive from Brussels, PV technology is growing in most countries. However, this growth is normally dependent on a national government market intervention and so is normally dependent on the finance minister maintaining the grant and incentive levels.

Solar PV panels used in urban areas

Solar PV panels used in rural areas

Solar PV panels can even be used on motor homes

How solar PV cell structure works

Solar PV cells are made from two wafer thin layers of semiconducting material, and frequently this is **silicon**. One layer is the positive or 'p' type, having been 'doped' with boron. The other layer, known as the negative or 'n'-type is made by **doping** with phosphorous. It is the junction between these two thin layers, the p–n junction, that is important.

When the Sun's energy, as **photons**, fall on the PV material near the p–n junction, an electron can gain enough energy to 'jump' out of its atom and move through the n-type **semiconductor**.

The electrons move slowly one way and the 'holes' they have created drift the opposite way. Another name for the p–n junction is the **depletion region** for this reason. The electrons move towards metal contacts on the front of the PV cell, resulting in a flow of electrons in the metal conductor, which is an electric current.

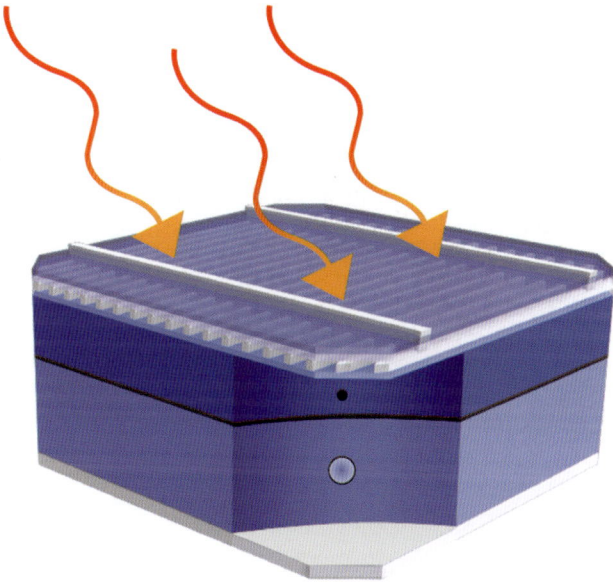

Illustration of a solar PV cell

The electrons flow in one direction, creating a direct or DC current. They flow into the metal strip on the top of the solar PV cell. The positive charged 'holes' 'migrate' to the lower side of the PV cell, which is covered with a metal plate, and eventually meet up with a **negative charged** electron and the situation is resolved, and the cycle can be repeated.

The difference between the two surfaces gives a small measurable **voltage** which drives the DC current.

Silicon A common semiconductor used in majority of PV cells

Doping Introduction of other materials or elements that improve the conversion efficiency of silicon

PV cell

Photon A single bundle of light energy

Semiconductor A substance with conducting properties partway between those of a conductor and an insulator

Depletion region An area within the silicon layers of a PV cell where photons of light release sufficient energy to cause electrons to flow, forming the electrical flow

Negative charge Electrons carry a very small negative charge

Voltage Voltage measures electrical difference between two points. It is the electromotive force that causes electric current to flow in an electric circuit. Symbol V

PV cell in a solar panel

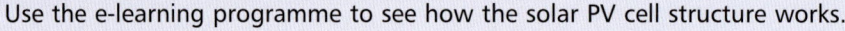

E-LEARNING

Use the e-learning programme to see how the solar PV cell structure works.

ACTIVITY 1

When it comes to electric shocks, what form of electricity, AC or DC, is potentially more dangerous?

Stand alone/ autonomous system A PV system not connected to the grid

Grid connected A source of renewable energy that is connected to the local distribution network

Stand alone vs grid connected systems

There are two different types of photovoltaic system:

● **Stand alone**

● **Grid connected**

Urban house with Solar PV panel installed

Stand alone system

Stand alone systems are used where there is no access to mains electricity or it is impractical to use it. In developing countries, the electricity grid tends to be confined to the main urban areas and does not reach rural areas. In this case it can be the best and least expensive way of providing electricity to these areas. In the developed world it is now common to see PV powered street signs, phone kiosks and weather monitors. You can even buy stand alone PV chargers for mobile phones and MP3 players.

Batteries are used as back up when it is dark, or demand exceeds production. The batteries are charged during PV panel operation.

Stand alone systems are best suited for areas with no mains electricity

Grid connected system

A mains connection means that the grid can provide additional electricity when there is insufficient sunlight to meet the electricity demand of the user. It also works in reverse in that excess electricity can be passed back via the grid to power other local houses and so power stations burn less fossil fuel. In Great Britain, grid connected PV and other renewable electricity gain both a generation tariff for all the electricity produced and an export tariff for all the electricity that is sold to the grid.

Example of a grid connected system

E-LEARNING

Use the e-learning programme to see a demonstration of the different connected systems.

ACTIVITY 2

Inverter A converter that transforms DC voltage and current into AC voltage and current

What would be the difference between an **inverter** installed in the USA and an inverter installed in Europe?

BENEFITS OF EFFICIENT SOLAR PV SYSTEMS

Efficient solar PV systems

The solar PV system needs to be designed to be as efficient and effective as possible. There are no moving parts within the system, so once installed it should require minimal attention and be trouble free.

Selection of the materials making up the types of PV panels needs to be right for the location. The size and correct positioning of the panels is crucial to maximizing the collection of solar energy, as is the installation with attention to good connections and equipment components. If it is a PV system connected to the National Grid, it will maximize the opportunity to sell the excess electricity to the local energy supplier.

National Grid, or Grid The high voltage electricity network designed to transmit electricity across the country from large power stations to regional distribution networks

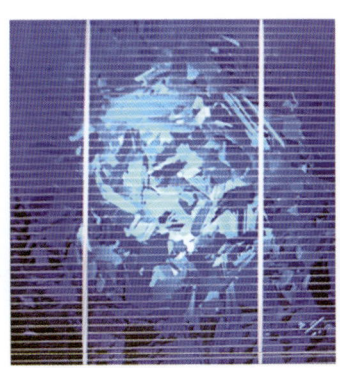

Solar photovoltaic (PV) panels

Solar PV performance

Solar PV systems emit no carbon dioxide and use a free energy source. Solar PV systems are also less site dependent than other renewables such as wind and hydroelectric.

Solar PV performance is affected by latitude, cloud cover, clarity of air, tilt, orientation and particularly shading.

Tilt The tilting of something away from a line or surface, or the degree to which it is tilted. For solar PV systems in the UK, the best angle of tilt or inclination is between 35° to 40° to the horizontal

Orientation Geographical direction the PV panels are facing. Optimum position is directly South. PV panels can work with other orientations but with loss of efficiency

Carbon footprint

ACTIVITY 3

There is an ever on going debate as to whether Solar PV is zero carbon or whether the 'embedded energy' is greater than the energy produced by the PV panel over its useful lifetime. What do you think about this debate?

Solar orientation

Inclination The tilting of something away from a line or surface, or the degree to which it is tilted. For solar PV systems in the UK, the best angle of tilt or inclination is between 35° to 40° to the horizontal

The panel orientation and angle of **inclination** towards the Sun can make a significant difference to their efficiency. The graph shows how this can be worked out.

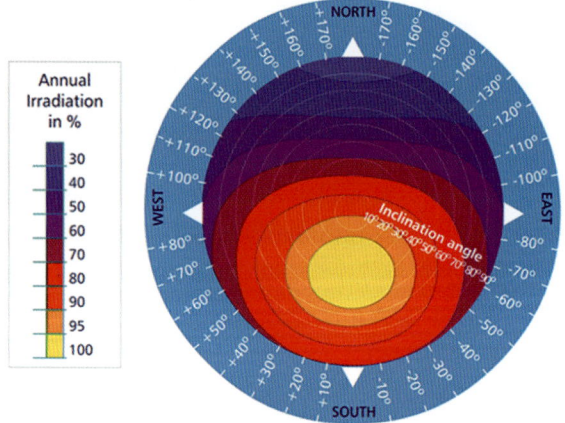

Graph illustrating how the panels inclination towards the sun effects efficiency

Orientation

In the UK, PV panels should ideally face south which receives most solar radiation. However, orientations between 35° east and 40° west of south are acceptable and will not lose more than 10 per cent in efficiency compared with the ideal situation. Alternatively the surface area of the panel could be increased to compensate for lower efficiency if outside these geographical positions.

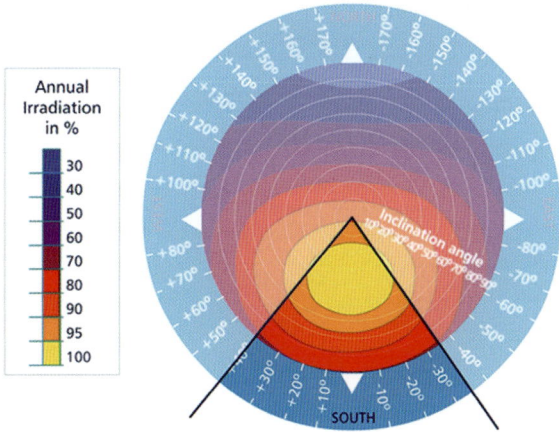

Graph illustrating how orientation of the panel effects efficiency

Angle of inclination

PV panels also need to be inclined towards the Sun and an angle of 35° to the horizontal is best for the UK. The graph shows there is some leeway in angles, depending on the pitch of the roof. As can be viewed on the graph, for a PV panel facing directly south, a panel at an angle of 5° or 65° will not lose more than 10 per cent as compared to the optimum 35°.

If the roof is facing east or west and so is not at the optimum angle, a solar PV system can still be fitted. In this situation, the panels should be 'oversized' to compensate for the loss in solar energy.

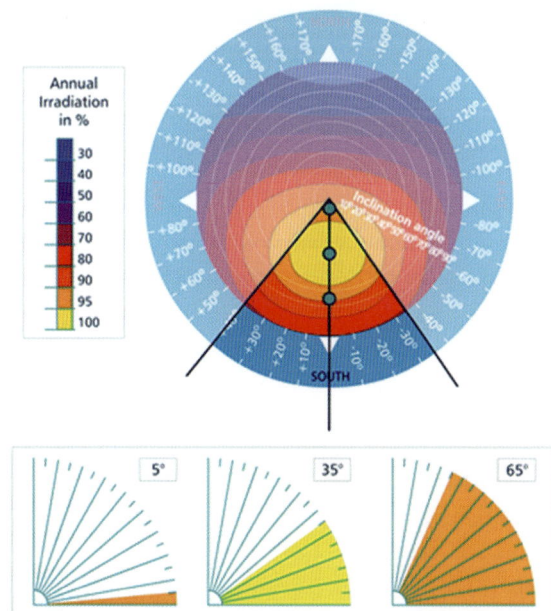

Graph illustrating the effects of the angle in inclination

E-LEARNING

Use the e-learning programme to see a demonstration of solar orientation.

ACTIVITY 4

In the above section, we have discussed the ideal angle and orientation for solar PV panels in the UK. How do you think this might change for other areas of Europe and how would it change again for the southern hemisphere?

RELEVANT REGULATIONS AND STANDARDS

Relevant regulations, standards and planning permission

UK AND INTERNATIONAL STANDARDS

Standards See British Standards

Installations of solar PV systems are covered by various building regulations and standards. These are particularly those concerned with the structural changes, energy efficiency, electrical equipment and electrical connections affected by the installation. Various standards also affect solar PV installations.

It is important that all relevant regulations, standards and guidance requirements are followed at all times when installing solar PV. Inspection, testing and commissioning must be carried out correctly.

HEALTH AND SAFETY

The main Building Regulations involved are Parts A Structure, B Fire Safety, L Conservation of Energy and P Electrical Safety.

By following the regulations the installation will be more safe and efficient, enabling more electricity to be added to any feedback tariff.

We have listed the English and Welsh Building Regulations here. They act as a good example of what happens in many European countries and similar regulations will probably be used in most European countries. All PV systems should be fitted with consideration of structural, safety, regulatory and energy conservation issues in mind and follow local requirements.

Main Building Regulations; parts A, B, L and P

Structure

Maintaining the structural integrity of the building is vital, as the modules, depending on their layout, can create wind uplift and so can cause structural damage. Fixing modules to the roof and walls also can affect the water-tightness of the building – it is important to prevent leaking roofs! PV installations can also affect the fire resistance integrity of the building.

PV installation can create wind uplift, affect water-tightness and fire resistance

Electrical

HEALTH AND SAFETY

Electrical safety procedures must be followed as solar PV cells can vary their output instantly as the Sun appears or is hidden by clouds or shadows.

The industry standard electrical wiring regulations must be followed as relevant to solar PV systems. This will also include system installation, inspection and testing, and commissioning.

Electrical safety procedures must be followed

Notification and energy efficiency

The system installed needs to be as efficient as possible, maximizing energy conservation.

UK AND INTERNATIONAL STANDARDS

Please note that the local building control will require notification of works under the relevant parts of the Building Regulations. This can be covered through self-certification or via a building control officer.

Notification under the relevant parts of the Building Regulations applies to England and Wales. Most countries will have their own procedures for notification or self-certification. Always make sure you follow the procedures relevant to the local jurisdiction.

Efficiency is key in maximizing energy conservation

Planning permission

Photovoltaic panels are counted as permitted development under the General Permitted Development Order and do not usually require planning permission. Exceptions apply if the panels extend more than 200mm over the existing roof plane, if the building is listed or if it is in a conservation area or world heritage site. If in doubt, it is always best to consult your local authority before beginning installation of photovoltaic panels.

Many people confuse the difference between Building Regulations and Planning Permission. Building Regulations cover the aspects of the building that must be complied with to make the system comply with the local construction requirements. Planning permission is related to the architectural aesthetics of the local environment. Most local authorities will have two separate approval systems for regulations and planning permission and it's important to make sure that all aspects of both regulations and planning are covered. Many authorities might be like England and Wales whereby they provide a presumption of consent to install renewable energy systems. However, if in doubt, always check local requirements.

If panels extend more 200mm over existing roof plane - planning permission needed

CHECK YOUR KNOWLEDGE

1. **Label the Solar PV cell shown.**

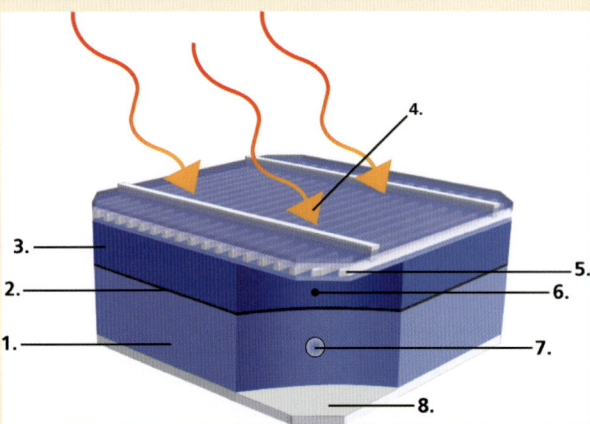

2. **Building regulations and planning permissions are relevant to solar PV installation. Complete the table below by circling when you need to notify or make an application.**

Building regulations	Always	Sometimes	Never
Planning permission	Always	Sometimes	Never

3. **Answer the following questions about solar radiation in the UK.**

    Circle the correct answer from the options shown:

What is the optimum angle for a solar module in the UK?	17°	35°	90°
Can solar PV modules use diffuse solar radiation?	Yes	No	Sometimes
What is the optimum direction for solar PV modules be facing in the UK?	East	South	West

Chapter 2

HEALTH AND SAFETY

LEARNING OBJECTIVES

By the end of this chapter you will be able to:

- List the key items of Personal Protective Equipment

- Identify common safety and hazard signs

- Describe the effects of hazardous substances

- Describe safe working at heights

- Identify the key elements of a first aid kit

- Know how to carry out a risk assessment

- Match fire extinguishers to different types of fire

- Select CSCS safety cards for different types of people

- Describe all aspects of electrical safety

HEALTH AND SAFETY AT WORK ACT

Health and Safety at Work Act (1974) (HASAWA) All employers are covered by the HASAWA, which places specific duties on both employers and employees to ensure the health and safety of everyone in the workplace

HASAWA 1974

The Health and Safety at Work Act, 1974 provides the legal framework to:

● promote

● stimulate

● encourage

high standards of health and safety in the workplace.

It protects employees and the public, putting the onus on everyone to be responsible for their own health and safety and of others who could be affected.

Before going any further though, it is essential that you have an understanding of the Health and Safety at Work Act, 1974.

The Act provides the legal framework to promote, stimulate and encourage high standards of health and safety in the workplace. It protects employees and the public from work activities.

Everyone has a duty to comply with the Act, including employers, employees, trainees, the self-employed, manufacturers and suppliers.

Must comply with HASAWA

HASAWA 1974

If negligence can be proved, both employers and employees can face a £5000 fine from a Magistrate's Court, unlimited fines and imprisonment from a Crown Court.

Employer responsibilities

The Act places a general duty on employers to ensure 'so far as is reasonably practicable the health, safety and welfare at work of all their employees'.

Employers must provide and maintain safety equipment and safe systems of work. This includes, among other things, ensuring materials are properly stored, handled, used and transported, providing information, training, instruction and supervision and ensuring staff are aware of manufacturers' and suppliers' instructions for equipment. Employers must also look after the health and safety of others, for example the public, and talk to safety representatives.

Employer

Employers are forbidden to charge employees for any measures which are required for health and safety, for example, personal protective equipment.

Employee responsibilities

Employees must comply with the Act too and look after their own health and safety as well as the health and safety of others. They must co-operate with their employers and not interfere with anything provided in the interest of health and safety.

Employee

PPE

'**PPE**' stands for 'Personal Protective Equipment' and covers a range of different items of clothing or equipment – such as gloves or safety helmets – that you may have to use on a job to avoid harm or injury.

Employers have a legal duty to identify any **risks** involved with a particular job and thus what items of PPE may be needed – but you still need to know the basics to ensure you are provided with what you need.

As ideally risks to health and safety should be eliminated from the workplace before they occur, PPE is a last line of defence. However, if risks remain, then PPE must be provided to you free of charge.

PPE Personal Protective Equipment

Risk Refers to how likely it is that a potential hazard will actually damage your health

Worker wearing PPE while at work

Common PPE items

You need to be familiar with several key items of PPE. You may not need to use them on every job, but you still need to know when they are required and how they should be used to carry out your daily work.

Solar panels being installed on the roof of a domestic dwelling

Safety footwear

For many jobs you will need to wear steel-toed boots with intersoles – thin pads in the boots or shoes that absorb surface shock – to protect your feet from injury.

If working in wet conditions, rubber boots should be worn – which must also have intersoles and steel toecaps.

Overalls

Safety footwear

Overalls

Boiler suits are ideal for many jobs as they provide cover for your entire body. However, you must never wear overalls made from terylene, nylon or similar materials, as these catch fire easily.

Ear muffs/ear plugs

If you are exposed to high noise levels, you must protect your hearing with ear muffs or ear plugs.

Ear muffs must always be properly fitted so that the ear is completely covered, otherwise you will not be fully protected.

Ear plugs fit inside the ear and are often disposable. They offer less protection than ear muffs.

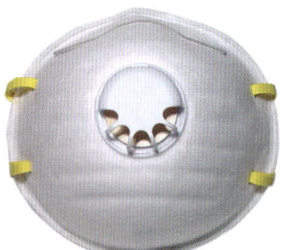

Ear muffs

Respirators

Many different types of respirator are available, but filter masks are the most common. These are rubber face masks that fit over the mouth and nose, containing a filter canister through which the wearer breathes.

Filter masks stop dust but are useless against gases or vapours – so if you are working around these, you must use a canister respirator. Filter canisters must be changed regularly.

Respirator

Safety gloves

Safety gloves protect your hands from injury. Take care to choose the right kind, as different types exist, for example for working with heat or chemicals. Check glove application data, if available.

Do not wear gloves when using machinery – e.g. drills – as this is dangerous.

Safety gloves

Safety goggles

You must wear safety goggles when required, for example when welding, working in dusty conditions, or when flying chippings will be produced.

Always check that you have the right type of goggles, as different lenses offer different levels of protection.

If goggles have dirty lenses, clean them before use: never obscure your vision.

Safety goggles

Safety helmet

Safety helmets

Hard hats should be worn on a building site whenever there is a risk of falling or flying objects. You must always check that your helmet fits properly. Always wear your hat with the peak facing forward, as the peak lip is designed to protect your eyes. When working indoors, a 'bump hat' can be worn instead.

Hard hats sometimes come with an expiry date. It is not advisable to use such hard hats after the expiry date as it may have deteriorated and not provide the necessary protection.

E-LEARNING

Use the e-learning programme to learn more about the different types of PPE.

SAFETY SIGNS

Work sites make extensive use of safety signs to warn of a number of different risks and hazards. You will learn what the different signs mean, but before you do it is important to understand what is meant by a 'risk' and a 'hazard' – as they are not the same thing.

Safety sign

Hazards

A hazard exists if a substance can potentially damage your health. The negative effects of a hazard may be relatively minor – such as making your eyes water – or they may be much more serious, such as suffocation.

In some cases, they can be fatal. Hazards may be difficult to detect, and need not affect you immediately, such as the cancers eventually caused by asbestos over the long term.

Risks

Risk refers to how likely it is that a potentially hazardous substance will actually damage your health. The better controlled a hazardous substance is, and the more rigorous people are about using PPE properly, the lower the risks will be.

Identifying safety signs

To be able to work safely you must understand what the basic safety signs mean.

'**Prohibition**' **signs**, which are circular with red crosses through them, tell you not to do something.

Prohibition sign
These are circular with red crosses through them, telling you **not** to do something

No smoking   *No lit matches*   *Do not extinguish with water*   *No pedestrians*

'**Mandatory**' **signs**, which are blue circles with white symbols, tell you what you must do.

Mandatory sign
These are blue circles with white symbols which tell you what you must do

Must wear safety goggles   *Must wear safety gloves*   *Must wear ear muffs*   *Must wear hard hat*

Most signs usually have accompanying textual information to further explain the sign.

'Warning' signs have yellow triangles with black symbols, these give notice of a particular hazard or danger.

Dangerous chemicals   *Danger acid*   *Danger*   *Irritant*

'**Information**' **signs** are used to communicate safety information. These are green squares with white symbols.

Information sign
These are green squares with white symbols to give information

Emergency phone   *Emergency exit direction*   *First aid*

CONTROL OF SUBSTANCES HAZARDOUS TO HEALTH (COSHH)

Classifying hazardous substances

Hazardous substances can be broken down into four main categories:

- Toxic/very toxic substances
- Corrosive substances
- Harmful substances
- Irritants

Being able to identify these in a practical setting will help to alert you to possible sources of danger so that you can take appropriate protective action.

Toxic/very toxic substances

Toxic/very toxic substances such as bleach can cause death or serious damage when inhaled, swallowed or absorbed by the skin.

Corrosive substances

Corrosive substances such as sulphuric acid may destroy parts of your body if they come into direct contact with them.

Harmful substances

Harmful substances such as lead can cause death or serious damage when inhaled, swallowed or absorbed by the skin.

Irritants

If irritants – such as soft solder flux – come into contact with the skin, eyes, nose or mouth, they can cause inflammation or swelling. Such effects may be felt immediately, or they may come after extended or repeated contact. Approach roofs onsite with caution.

A warning sign in use on a building site

UK AND INTERNATIONAL STANDARDS

Under European law, substances are defined as being officially hazardous to health if they are listed as being 'dangerous to supply'.

They must also be defined as being very toxic, toxic, corrosive, harmful, or irritants. One practical result of this is that when sold commercially, the packaging for such substances must be clearly marked and labelled.

Hazard signs

Hazardous substances are identified by four main **warning signs**.

Toxic/very toxic
substances

Harmful substances

Irritant

Corrosive substances

Warning sign These warning signs are yellow triangles with black symbols and give notice of a particular hazard or danger

Corrosive material

Substances marked as being 'corrosive' could cause permanent damage if they come into direct contact with any part of your body. For example, sulphuric acid will burn your skin away and cause breathing problems.

Flammable material

Flammable materials must be kept away from naked flames, and you should not smoke when near them. Common flammable materials include your own clothes, hair that is worn long, some modern hair products and any oily rags that may have been left lying around.

Explosive material

Explosive materials must be handled and stored with particular care, as they potentially present an extreme hazard.

Toxic material

Some toxic materials, such as gas in a confined area, can harm you even if you do not come into direct contact with them, so ensure you handle them with care. If necessary, seek advice as to whether PPE is required when they are in use.

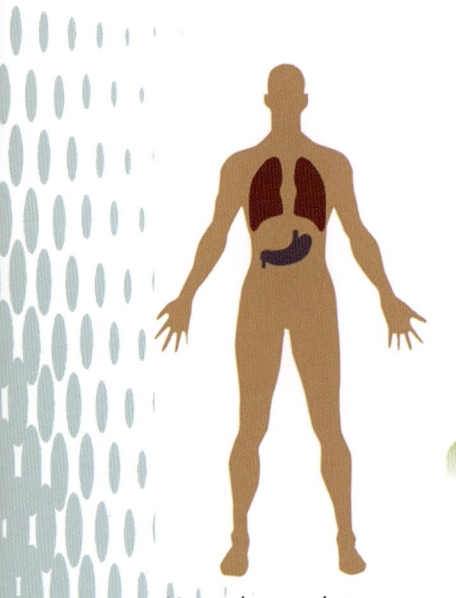

Hazardous substances can damage skin, eyes, lungs and stomach

The effects of hazardous substances

In order to be able to adequately follow safety procedures and use PPE correctly, it is important to understand exactly how harmful substances can affect your body.

Skin and eyes

Some substances will cause damage – such as burns – if they come into direct contact with the skin. Contact may also lead to less serious problems, such as increased skin sensitivity. Harmful substances such as solvents can also enter the bloodstream by being absorbed through the skin, or through cuts and bruises.

Lungs

Harmful substances can cause damage when inhaled in one of two ways. Either they can cause direct damage to the lungs themselves in the way that asbestos does, or they can enter the bloodstream and so affect other organs.

Stomach

Harmful substances such as lead can reach the stomach in numerous ways if basic hygiene is not observed, or if gloves are not worn. Eating, drinking, smoking and biting your nails can all be responsible for this.

E-LEARNING

Use the e-learning programme to learn more about the effects of hazardous substances.

WORKING AT HEIGHTS

Using ladders safely

People injure themselves using ladders for four main reasons. They over-reach or slip, or the ladder itself breaks or falls over. These risks can be minimized if you stay aware of how to use ladders properly.

Ladders

Leaning ladders

Before you use a leaning ladder, check that it has clean rungs and undamaged stiles (these are the side pieces the rungs are attached to). Position it at an angle of 75° from the vertical (e.g. 1m out from the base for every 4m of height).

Ensure that it can't move about at the top, that it has a strong upper resting point and that its rungs are horizontal. When in use, never stand on the top three rungs and maintain three points of contact at all times.

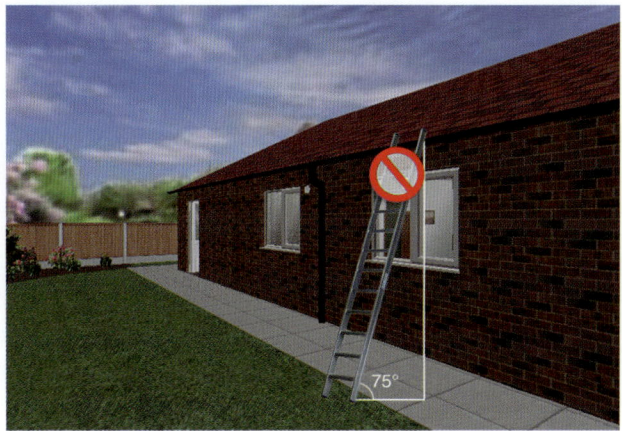

Leaning ladder must be angled at 75 degrees

Step ladders

Before you get started, check that the stepladder is in good condition, with clean treads and secure locks. It should be fully open and locked firmly in place. Do not work sideways on.

Step ladder in use

Before you start – general principles

Whatever type of ladder you are using, before you start you should check that it is in good general condition. Are its feet firmly attached? Have you properly secured its fastenings?

You must also ensure that it is in a good position. It should not be able to move at the bottom and should be placed on a firm, level surface that is clear and dry.

Finally, you must always be fully fit to work at heights, and never paint wooden ladders as this may conceal damage or cracks.

When using ladders, they must be in a good working condition.

Using ladders – general principles

You should never work on a ladder for longer than 30 minutes and must not exceed stated weight limits, so only carry light tools and materials.

When working, keep your body centred on the ladder and avoid over-reaching. You must also keep both feet on the same rung or step and when you climb, always keep a firm grip on the ladder.

Non-slip footwear should always be worn.

General principles to follow when using ladders

Safety boots must be worn

Safety gloves must be worn

Do not work on a ladder for longer than 30 minutes

Using scaffolding safely

Setting up, using and checking scaffolding must all be done with great care, otherwise falls and collapses can cause serious injury, or even death. Also, due consideration must be paid to public safety.

Great care must be taken when using scaffolding to avoid serious injury

Erecting scaffolding

Scaffolding must always be erected by competent people. It should be put up on firm, level ground away from any power lines. Uprights should be placed on base-plates.

Gaps are dangerous, so it should be no more than 300mm away from the building or structure it is attached to – and such attachments must be secure.

Working platforms must always be wide enough for people to use equipment and pass each other in safety – no less than 600mm wide and free of openings.

Using scaffolding

To avoid falls, strong barriers such as guardrails and toe boards must be in place. Heavy or bulky loads should never be carried up or down ladders, which should be strong, secure and in good condition. The top of the ladder should project 1m above the level of the work platform.

The use of domestic ladders should be avoided, as these may collapse. Under no circumstances should components ever be removed from the scaffolding after it has been set up.

Risk to the public

Scaffolding should never normally be set up over busy public areas. If a risk exists, limit work to quiet times. Prevent waste materials from falling on people by keeping work platforms clear of debris – if they are allowed to build up, they may become stacked above the level of the toe boards.

Always avoid using scaffolding over a busy public area.

Checking

Once it has been set up, scaffolding should be checked on a weekly basis, as well as after any alterations, damage or extreme weather. Such inspections must be done by a competent person. Any problems should be fixed immediately, or the scaffolding must be taken out of use.

Scaffolding needs to be checked on a weekly basis

Tower scaffolds

Before use, you must ensure the tower is braced, has outriggers and that any wheels are locked. Tower scaffolds must be kept away from overhead **cables** and ladders should never be leant against them. You should never move a tower when people or heavy items of equipment are on board. Finally, as with ladders, you should avoid over-reaching when you are standing on one.

Ensure tower is braced, has outriggers and wheels are locked

Cable A conductor used to carry current around an installation. Cables are identified by the colour of the installation. DC cabling is also heavy duty and usually Class II

E-LEARNING

Use the e-learning programme to learn more about using scaffolding safely.

BASIC FIRST AID

However well prepared people are, accidents will happen. Employers have a legal duty to ensure that first aid kits are in place, containing sufficient supplies for all workers onsite.

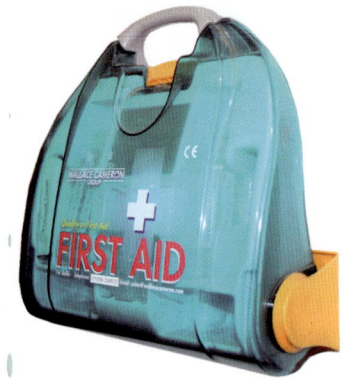

First aid kit

For smaller work sites, clearly marked first aid boxes must be placed under the control of a named individual. However, for larger sites of over 50 people, there must also be at least one person with first aid training.

MANUAL HANDLING

Manual handling includes lifting, carrying, lowering, pushing and pulling. We will consider lifting first, which can be broken down into five steps. The first is Stop and Think, when you plan the lift and assess the risks.

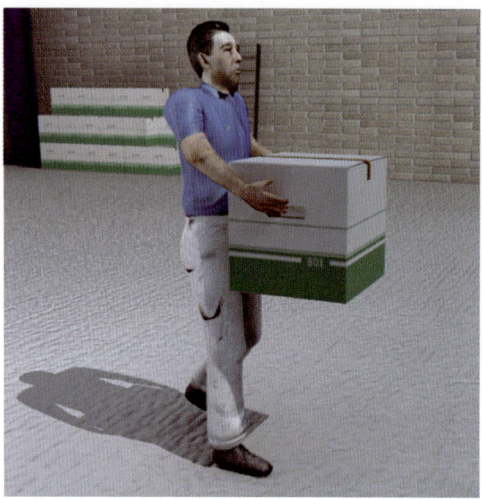

Carrying

Lifting

1. Stop and think – plan the lift and assess the risks

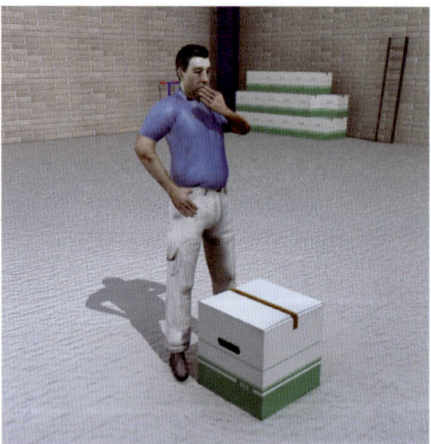

Step 1

2. Take up a good posture to start the lift and keep your back naturally straight

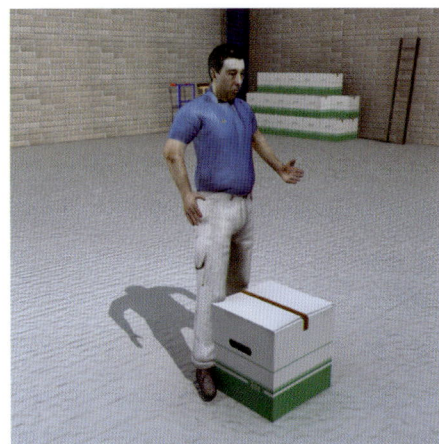

Step 2

3. Take a firm grip on the load, no twisting or over-reaching

Step 3

4. Start to lift using power of the legs to make a smooth and controlled action. If the load is too heavy do not continue

Step 4

5. Keep the load under control, heaviest side closest to the body, and beware of liquids and uneven loads

Step 5

Carrying

Carrying the load is the main purpose of manual handling. When carrying the load move your feet steadily and slowly, look ahead (not at the load) and avoid twisting the body.

● Keep the load close to maintain control
● Be vigilant and aware of your surroundings – especially be alert to possible danger, in addition to original assessment of risks. Be ready for the unexpected
● If feeling tired or strained, stop. Don't overdo it.

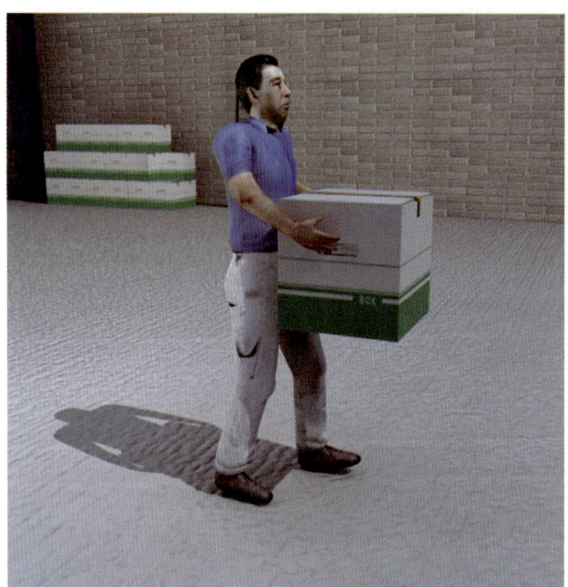

Carrying the load

Pushing

For loads that are too large to be carried by one or two people, an alternative must be found. Options available include trolleys, wheel-barrows and scissor lifts.

- Always check the trolley or barrow is in good working order
- Assess the whole route for hazards, especially slopes up or down and uneven surfaces
- Use safe lifting techniques to load the trolley
- Take a firm grip on the handles, take the weight of the load and lean forward slightly, keeping a natural straight back
- Use the power of the legs to push in an even and controlled way, about slow walking pace
- Once moving keep a steady pace and keep feet away from the load
- When you arrive at the destination slow down smoothly (no jerking movements) and unload in the same way it was loaded
- Always return the manual handling aid to its place to avoid turning it into a hazard, or preventing someone finding it when needed. If the aid cannot be found they may be tempted to do unsafe manual handling.

Use trollies, wheelbarrows or scissor lifts when the load is too heavy

Pulling

When you have a load on a trolley, or other manual handling aid, you can push or pull it. Pulling is more dangerous than pushing as the load can 'run over' you. If you lose control when pushing, the load will move away from you, however if you were pulling, the load will run towards you! Pulling also puts more strain on the back and it is less easy to control the load.

- Always check the trolley or barrow is in good working order
- Assess the whole route for hazards, especially slopes up or down and uneven surfaces
- Use safe lifting techniques to load the trolley
- Take a firm grip on the handles, take the weight of the load and lean backward slightly, keeping a natural straight back
- Maintain a steady pace – about slow walking pace and keep feet away from the load
- When you arrive at the destination slow down smoothly (no jerking movements) and unload in the same way it was loaded
- If uncomfortable at any time stop and reassess the situation. You may need to lighten the load or get help.

Pulling is more dangerous than pushing

Pointers to good manual handling practice

Carrying the load is the main purpose of manual handling. When carrying a load move your feet steadily and slowly, look ahead and not at the load and avoid twisting the body. Keep the load close to maintain control – especially if the load is a difficult one such as a liquid or odd shape. Be vigilant and aware of your surroundings and especially be alert to possible danger. Your original assessment of risks was taken at a particular time – things may have changed, so be ready for the unexpected! Remember if you are feeling tired or strained – stop. Don't overdo it.

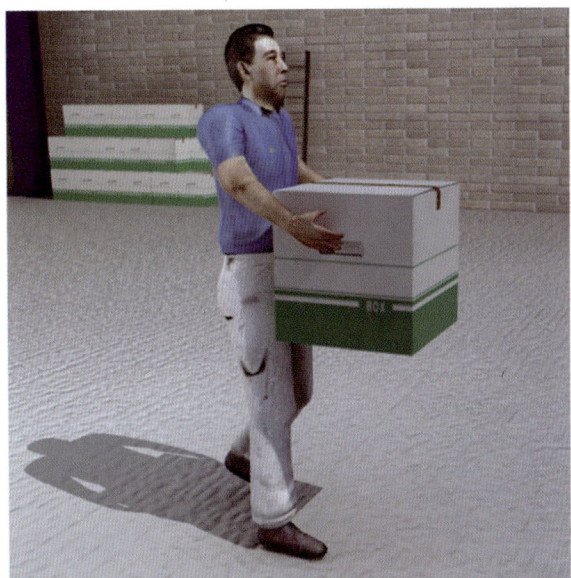

When carrying the load maintain good posture

Assessing health and safety risks in the workplace

Many accidents in the workplace should simply not happen. This is why it's very important to assess risks in the workplace, in order to protect not only the people working there, but also members of the public.

There is a legal requirement to carry out regular **risk assessments**, and, although the law does not expect all risks to be eliminated, people must be protected as far as it is reasonably practicable to do so. It is very important to regularly assess equipment in the workplace. The equipment shown below and on the following page are some of the key items that must be regularly risk assessed.

Risk assessment
Identifying hazards in the workplace then deciding who might be harmed and how

It is very important to assess the risks at the workplace

Tool box

High voltage

Leaning ladder

Tower scaffold

Keep a record of risk assessments

What is risk assessment?

Risk assessment means identifying hazards in the workplace then deciding who might be harmed and how.

A hazard can be anything that might cause harm, for example, electricity, chemicals, working from ladders or an open drawer.

There are five steps in the risk assessment process. Once this process has been completed, suitable precautions can be put in place to reduce the risk of harm, or make the harm less serious.

The five steps of risk assessment are:

● Identifying the hazards

● Identifying who could be affected

● Evaluating the risks

● Record findings and implement them

● Review risk assessment

Ladders can be considered a hazard

Step 1 – identify the hazards

The first step is to identify what the hazards are. When you work in a place every day, it's easy to overlook potential hazards, so there are ways to make sure you identify those that matter.

Walking around to look for things that might reasonably be expected to cause harm is a good starting point. Also, ask employees, or their representatives, as they might be aware of things that are not immediately obvious to you.

Publications and practical guidance are available from a variety of sources, including the Health and Safety Executive (or HSE), relevant trade associations, as well as manufacturers' instructions. These all provide information on where hazards can occur, their harmful effects and how to control them.

Accident and ill-health records are another source of information which can often help to identify the less obvious hazards. Don't forget the long-term health hazards, for example, those that can occur following prolonged exposure to high levels of noise or harmful substances.

Here are some hazards:

- Trailing cables might cause people to trip
- Spills might cause people to slip and also be toxic

Spills can cause people to slip

- Faulty or damaged electrical fittings might cause electric shock and possibly a fire

Faulty/damaged electrical fittings can be dangerous

- Items left lying about might cause people to trip or block access

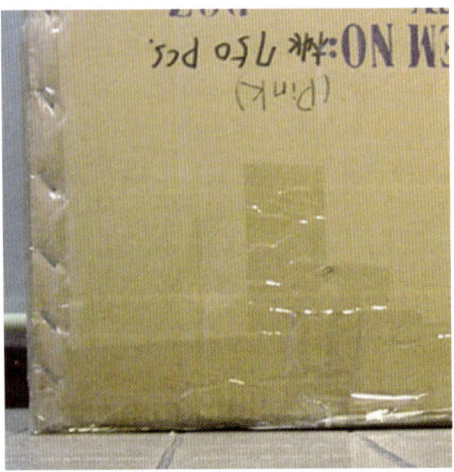

Items left lying about could cause accidents

● Lifting heavy loads might cause back injuries

Incorrect manual handling

● Long-term exposure to excessive noise might cause loss of hearing

Exposure to excessive noise may cause hearing loss

Step 2 – Who might be harmed and how

The second step in risk assessment starts with deciding who might be harmed by each of the identified hazards.

To do this, it's best to identify groups of people, rather than listing people by name, as this will help later on when it comes to identifying the best ways to manage the risk for each group.

It's also necessary to identify how the different groups of people might be harmed, for example, what type of injury, or ill-health might occur. Here are some people who have different requirements:

- Customers or members of the public who are visiting the workplace

Customers or members of the public

- Contractors who might not be in the workplace all the time

Contractors

- Maintenance workers visiting the site

Maintenance worker

● People with disabilities have their own requirements

A gentleman in an electric wheelchair

● Storemen responsible for moving heavy boxes

Storemen

● Expectant women might be at particular risk

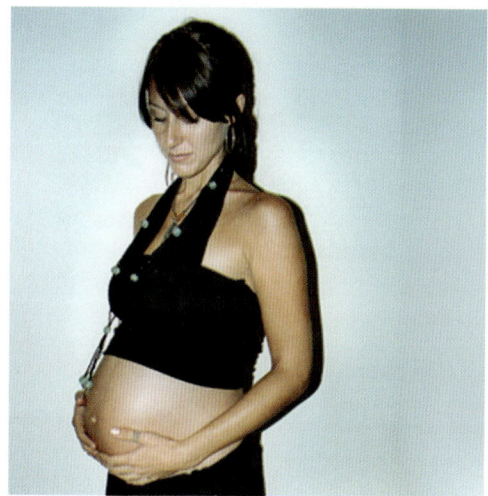

Pregnant women

Step 3 – Evaluate the risks and decide on precautions

Having spotted the hazards and worked out which groups of people might be affected by them and how, the next step is to decide what needs to be done to protect people from harm.

The law requires that everything 'reasonably practicable' is done, so the easiest way to find out if this is the case is to compare what is already being done with good practice.

FUNCTIONAL SKILLS

Again, the HSE website and HSE Infoline are sources of information about good practice.

It is important to evaluate the risks and decide what precautions to take

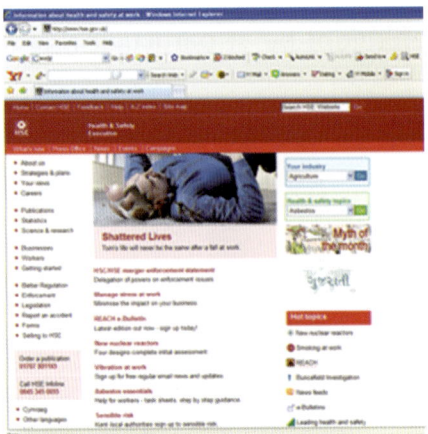

HSE Website

The first thing to consider is whether or not the hazard can be eliminated. If not, there are certain actions that should be applied, in the order shown, to control the risk and making sure harm is unlikely.

Improving health and safety doesn't need to cost a lot and failure to take simple precautions could cost a lot more if an accident occurs. Before introducing new precautions, always check that these are reasonable and do not introduce any new hazards.

Health and safety checks being carried out on a building site

Here are some different actions that could be applied:

- Switch to using a less hazardous method
- Switch to using a less hazardous chemical
- Prevent access by guarding the hazard
- Provide lifting equipment
- Provide clothing, footwear, goggles, etc.
- Provide first aid and washing facilities for removal of contamination.

Step 4 – Record findings and implement them

After you've spotted the hazards, worked out which groups of people might be affected by them and how, plus decided on measures to protect people from harm, it's important to keep a record of what's been done.

In fact, for businesses with five or more employees, the results of the risk assessment must be written down and actions recorded as they are implemented.

Smaller businesses will also find it useful to have a written record of their risk assessment, as it can be reviewed at a later date.

The main thing is to keep the written results of the risk assessment as simple as possible, and to share the document with employees.

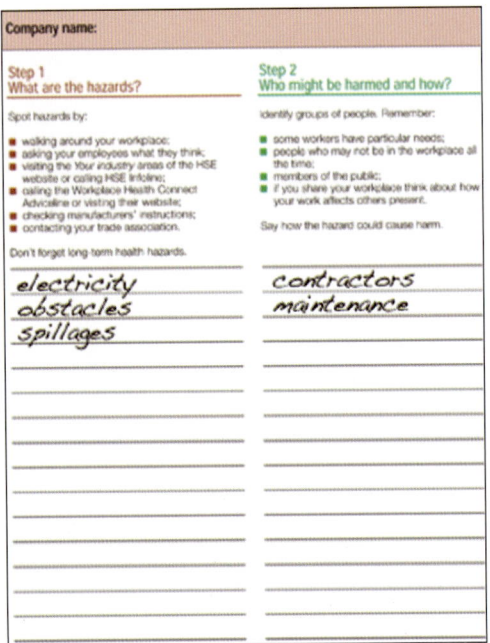

Company name:

Step 1
What are the hazards?

Spot hazards by:

- walking around your workplace;
- asking your employees what they think;
- visiting the Your industry areas of the HSE website or calling HSE Infoline;
- calling the Workplace Health Connect Adviceline or visiting their website;
- checking manufacturers' instructions;
- contacting your trade association.

Don't forget long term health hazards.

electricity
obstacles
spillages

Step 2
Who might be harmed and how?

Identify groups of people. Remember:

- some workers have particular needs;
- people who may not be in the workplace all the time;
- members of the public;
- if you share your workplace think about how your work affects others present.

Say how the hazard could cause harm.

contractors
maintenance

Recording findings

If quite a lot of improvements need to be made, it's best not to try and do everything at once, but to draw up an action plan to deal with the most important ones first.

A good action plan will often include implementing a few cheap or easy improvements which can be done quickly, perhaps as a temporary solution until more reliable controls can be put in place.

The plan of action might also include long-term solutions for those risks which are most likely to cause accidents or ill-health, or which have the worst potential consequences, as well as the arrangements which are made for training employees on how the remaining risks will be controlled.

And finally, the action plan should include who has responsibility for the various actions, and by when, as well as how regular checks will be made, to make sure that control measures stay in place.

Date of risk assessment:		
Step 3 What are you already doing?	**Step 4** How will you put the assessment into action?	
List what is already in place to reduce the likelihood of harm or make any harm less serious.	You need to make sure that you have reduced risks 'so far as is reasonably practicable'. An easy way of doing this is to compare what you are already doing with good practice. If there is a difference, list what needs to be done.	Remember to prioritise. Deal with those hazards that are high-risk and have serious consequences first. Action Action Done by whom by when

Plan of actions

Step 5 – Review risk assessment and update if necessary

Having identified the hazards, who might be harmed by them and how, what the risks are, the necessary measures and kept a record of what's been done, the fifth step in the risk assessment process is to review the risk controls and to update them as necessary.

It's a good idea to do this on an ongoing basis by thinking about risk assessment when changes are being planned, as well as conducting a formal, annual review.

Reviewing the risk assessment regularly will mean that controls can be amended each time new hazards are introduced, for example, when there are significant changes in the workplace, with the introduction of new equipment, substances or procedures, when problems are spotted by employees, or when accidents or near misses occur.

By carrying out regular reviews, you can ensure that risk controls are always up to date and improving and not sliding back.

Fifth step – review risk assessment and update if necessary

Responsibilities

Both employers and employees have responsibilities for health and safety in the workplace.

Employers are responsible for ensuring risk assessments are carried out on a regular basis. The process doesn't need to be a complicated one, nor does it need a health and safety expert to do it.

Employees have a responsibility to co-operate with their employer's efforts to improve health and safety, by complying with the controls which are in place, and by looking out for each other.

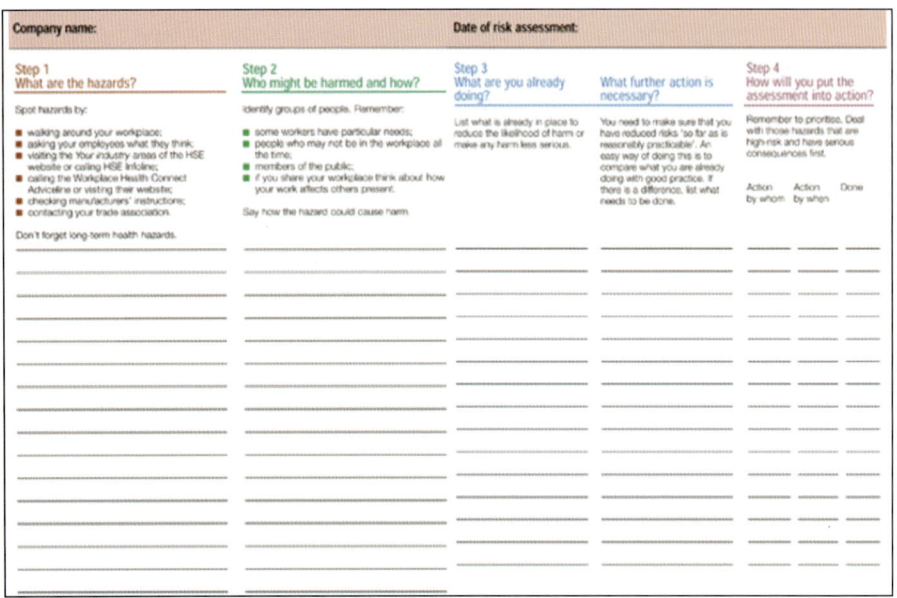

Risk assessment document

FIRE PROTECTION

Classes of fire

There are four common classes of fire:

- Class A: Solids – wood, paper, textiles, etc.
- Class B: Flammable liquids – oil, petrol, paint, etc.
- Class C: Flammable gases – acetylene, propane, butane, etc.
- Class D: Metals – magnesium, aluminium, sodium, etc.

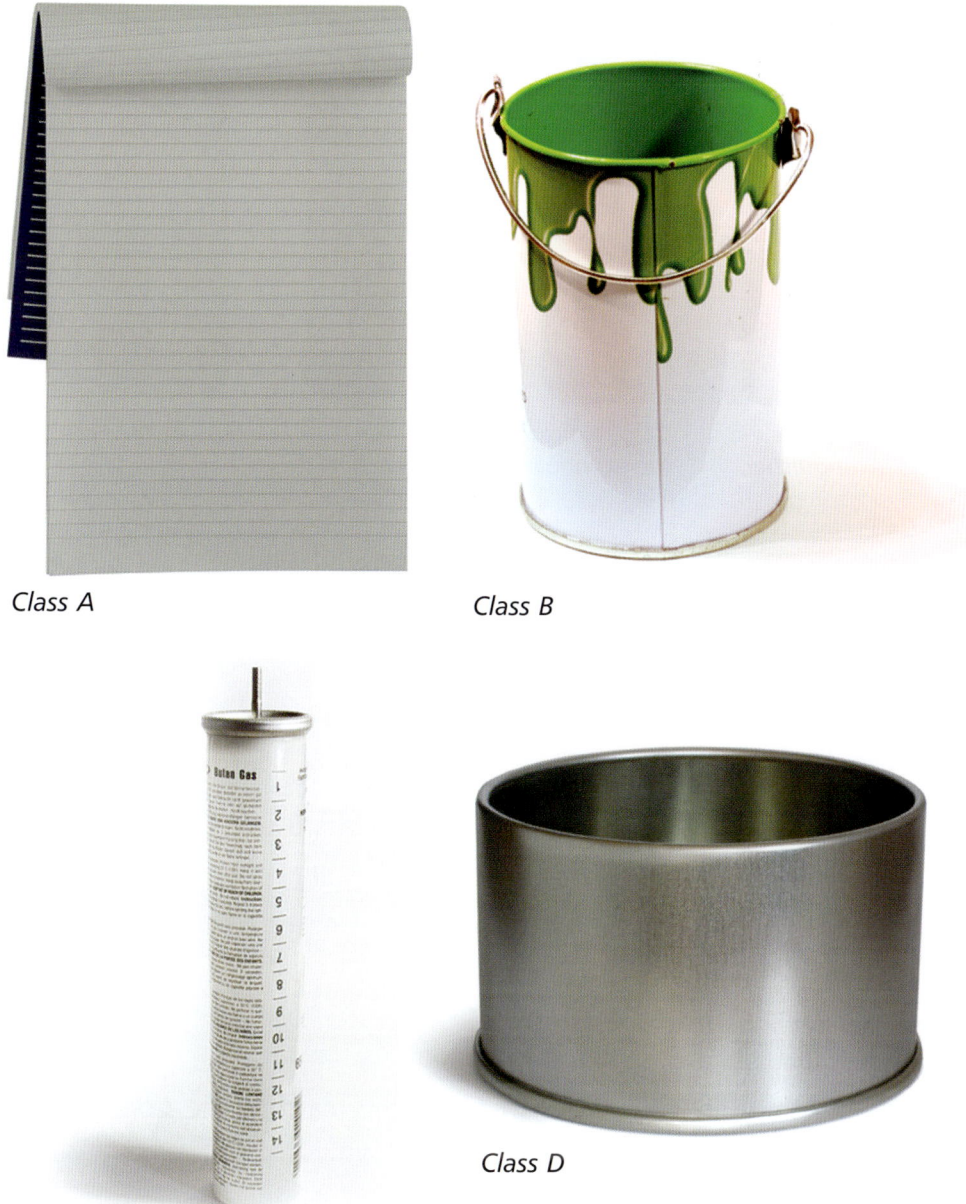

Class A

Class B

Class C

Class D

The type of burning materials tells you which type of fire extinguisher to use.

As electrical fires do not fall into any particular 'class' of fire, if an electric spark ignites, for example, paper, you would use a Class A extinguisher.

Types of fire extinguisher

There are four different types of fire extinguisher, which are shown below.

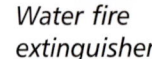

Water fire extinguisher   *Carbon dioxide fire extinguisher*   *Foam fire extinguisher*   *Powder fire extinguisher*

All fire extinguishers are now red, and are labelled to identify which type is which. It's important to select the correct one for the class of fire, otherwise it could have serious consequences.

- A water fire extinguisher is suitable for class A fires
- A carbon dioxide fire extinguisher is suitable for class B, C and D fires
- A foam fire extinguisher is suitable for class B and D fires
- A powder fire extinguisher is suitable for all classes of fire (A, B, C and D)

CSCS SITE SAFETY CARDS

There are nine different types of site safety card in the Construction Skills Certification Scheme, or CSCS as shown.

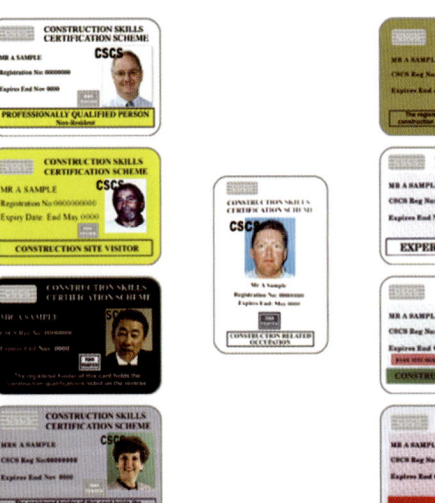
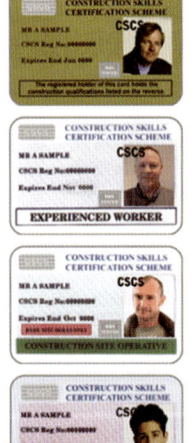

CSCS safety cards

Each card is issued to people in relation to their relevant experience and qualifications, and requires completion of the appropriate health and safety course or test.

Here are the details of the types of people who qualify for each of the two red cards.

Red Card – Trainee (Craft and Operative): registered for NVQ or SVQ (or Construction Award) and not yet reached Level 2 or 3.

Red Card – Trainee (Technical, Supervisory and Management): registered with a further/higher education college for a nationally-recognized, construction-related qualification or satisfactorily completed such a course.

Here are the details of who qualifies for a green card.

Green Card – Construction Site Operative: who carries out basic site skills with Level 1 NVQ or employer's recommendation using industry accreditation.

Here are the details of the types of people who qualify for each of the three blue cards.

Blue Card – Experienced Worker: with at least one year's on-the-job experience in last three years who missed opportunity for industry accreditation. Card is valid for one year whilst achieving Level 2 or higher NVQ/SVQ but is not renewable.

Blue Card – Skilled Worker: with Level 2 or higher NVQ/SVQ or completed employer sponsored apprenticeship or completed City & Guilds of London Institute Craft Certificate.

Blue Card – Experienced Manager: with at least one year's on-the-job experience in last three years. Card is valid for three years whilst achieving Level 3 or higher NVQ/SVQ but is not renewable.

Here are the details of who qualifies for a gold card.

Gold Card – Skilled Worker: with Level 3 NVQ/SVQ or approved indentured apprenticeship or employer sponsored apprenticeship and completed City & Guilds of London Institute Advanced Craft Certificate.

Gold Card – Supervisor: with Level 3 NVQ/SVQ in supervisory occupation or industry accredited.

Here are the qualifications required for a platinum card.

> **Platinum Card – Manager:** with industry accreditation or Level 4 NVQ/SVQ.

Here are the qualifications required to obtain a black card.

> **Black Card – Senior Manager:** with Level 5 NVQ/SVQ.

Here are the details of the type of people who qualify for a yellow card.

> **Yellow Card – Professionally Qualified Person:** consultants who are chartered members of approved institutions (e.g. architects, surveyors and engineers) with health and safety responsibilities and on site no more than 30 days in six month period.

Here are the details of who is eligible for a white card.

> **White Card – Construction Related Occupation:** for those occupations not covered by other cards.

Here are the details of the yellow visitor card.

> **Yellow Card – Regular Visitor:** with no specific construction skills who often visits a construction site. Makes site access easier as the holder would have passed a health and safety test before the card was issued.

ELECTRICAL SAFETY

Looking after and maintaining electrical equipment

The bottom line is that all electrical equipment should be looked after and maintained in a safe condition. But this doesn't always happen, as the pictures on page 57 show.

Continuously moving extension leads makes them prone to being damaged. If any damage occurs they should be replaced.

Extension leads that are moved a great deal are particularly prone to being damaged.

If the cable, plug or socket is damaged they should be replaced.

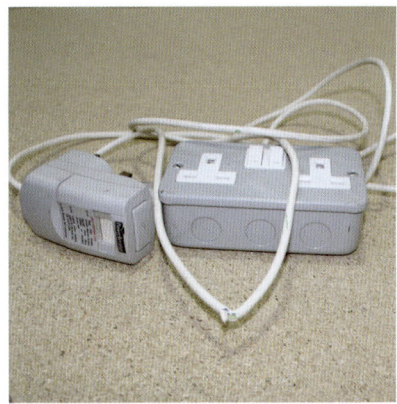

Extension leads that are often moved are prone to damage

Plug cables should always be firmly clamped.

The outer sheath of flexible cables must always be firmly clamped to stop the wires (particularly the earth) from pulling out of the terminals.

Plug cables should be firmly clamped

Always use proper connectors to join cables.

Cables should always be joined with proper connectors or cable couplers, not with strip connectors and insulating tape.

Always use proper connectors to join cables

Lamps and equipment which can easily be damaged must be protected to prevent risk of electric shock.

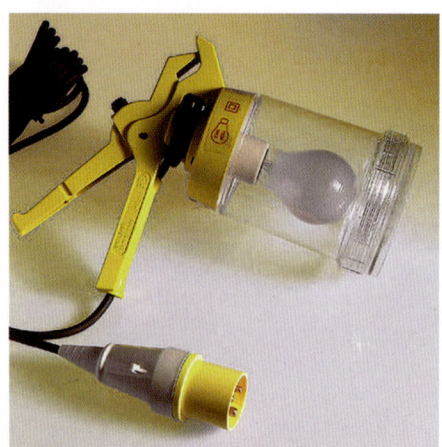

Equipment that can be easily damaged must be protected

Suspect or faulty electrical equipment must be labelled 'DO NOT USE' and kept secure until it can be examined by a competent person.

Equipment unsuited for use in a wet or harsh environment can easily become live and also make the surroundings live.

Suspect or faulty electrical equipment must be labelled "Do not use"

Some equipment is not suitable for use in wet or harsh enviroments

Pre-visual inspection

Many faults with electrically operated power tools can be found by visual inspection. By following a simple process before using the equipment, you can minimize most electrical risks, as shown here:

- Switch off and unplug
- Check plug is correctly wired

- Check fuse is correctly rated by checking equipment rating plate or instruction book
- Check plug is not damaged, cable is properly secured and no internal wires are visible
- Check cable is not damaged and has not been repaired with insulating tape or unsuitable connector
- Check outer cover of equipment is not damaged, which might give rise to electrical or mechanical hazards
- Check equipment for burn marks or staining that might suggest equipment is overheating.

Switch off

Unplug equipment

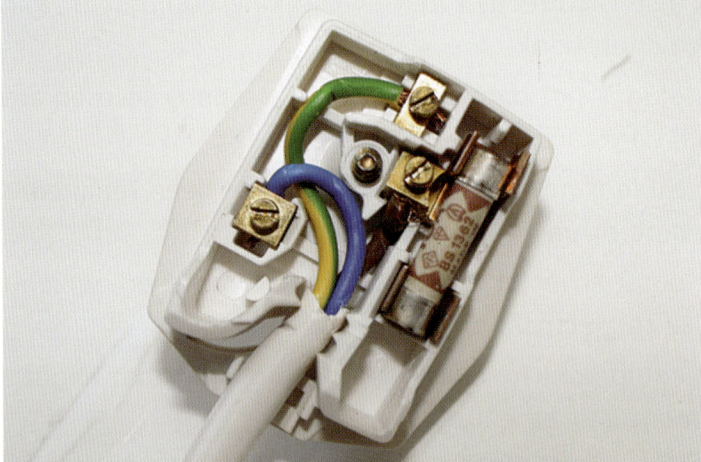

Check plug is correctly wired

Check fuse is correctly rated

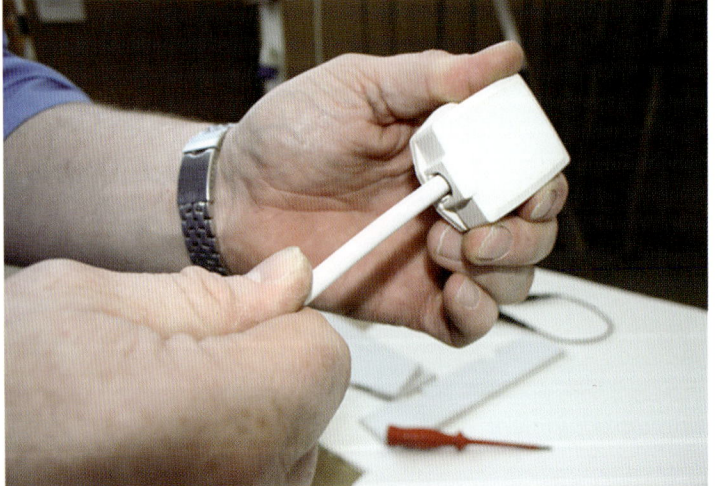

Check plug is not damaged

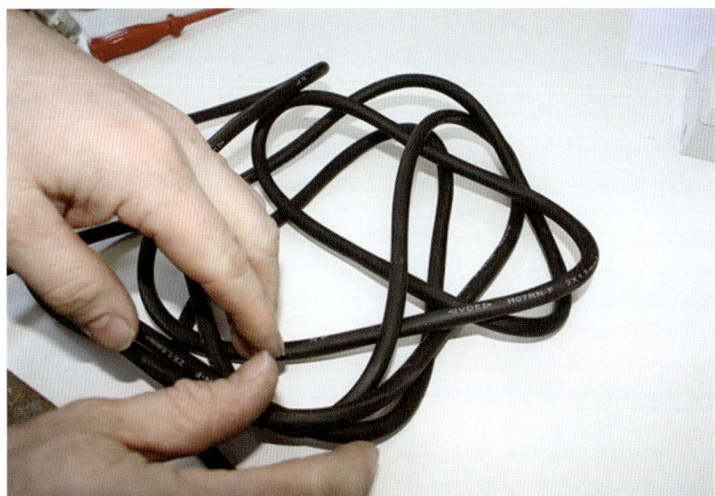

Check cable is not damaged

Check outer cover of equipment is not damaged

Check equipment for burn marks or staining

E-LEARNING

Use the e-learning programme to see a demonstration of a pre-visual inspection.

Electrical site safety

UK AND INTERNATIONAL STANDARDS

When it comes to the safe isolation of electrical supplies and energizing electrical installations, it is also important to comply with statutory health and safety requirements, as laid down by the Electricity at Work Regulations.

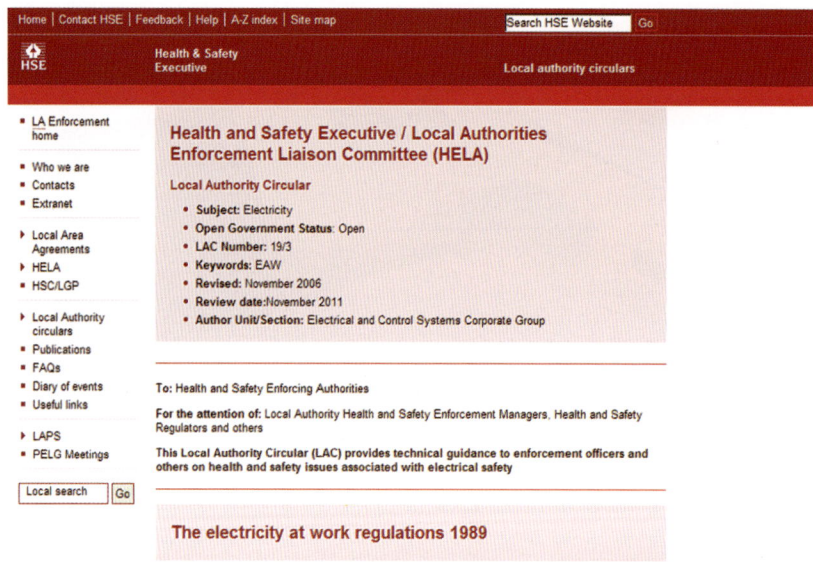

HSE Webstie

Safe isolation of electrical supplies

In order to avoid fatal accidents, which can occur during the proving of isolation, you should follow the recognized procedure:

- Identify source of supply
- Identify type of supply
- Isolate
- Secure the isolation
- Test the equipment/system is dead
- Begin work.

Step 1: Identify the source of supply

Step 2: Identify the type of supply

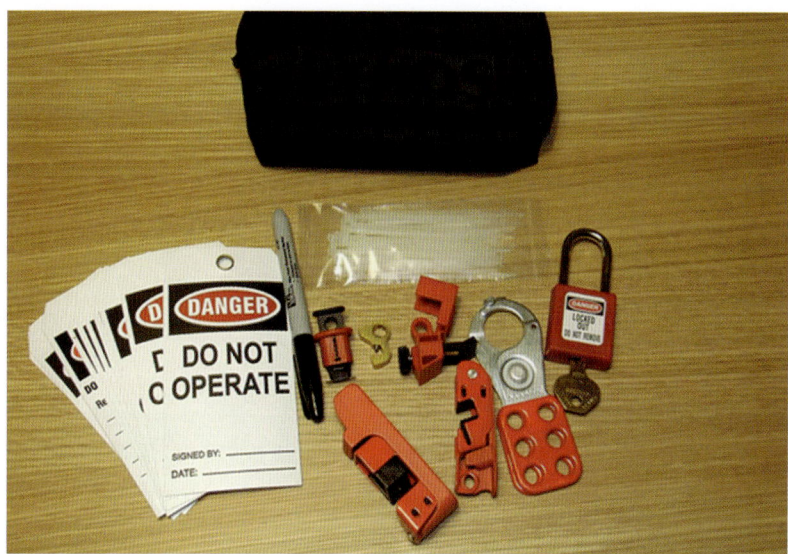

Step 3: Isolate

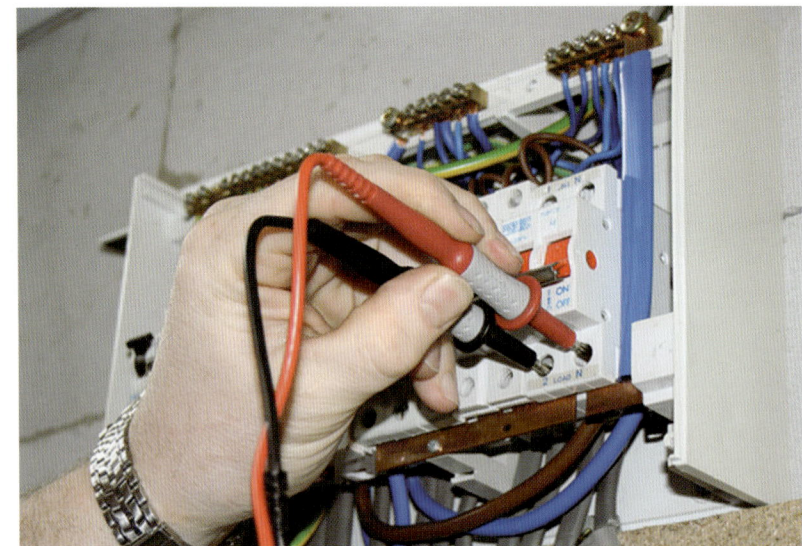

Step 4: Secure the isolation

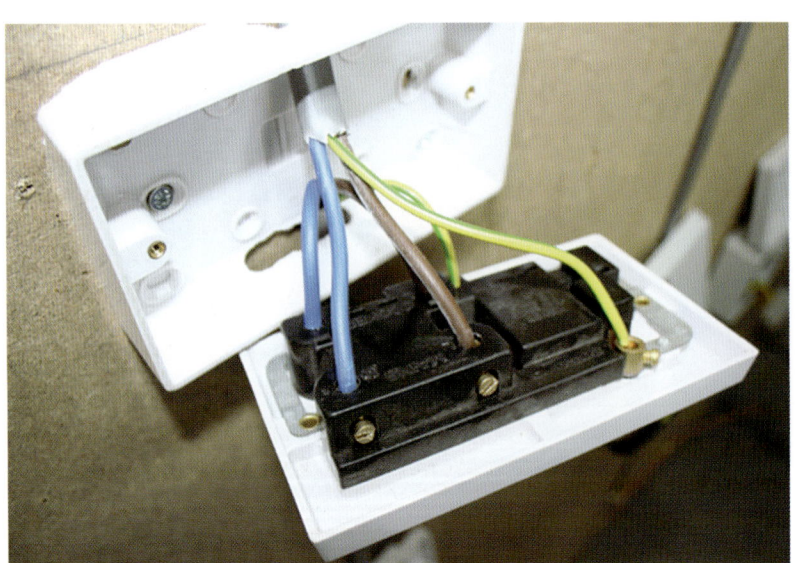

Step 5: Test the equipment/system is dead

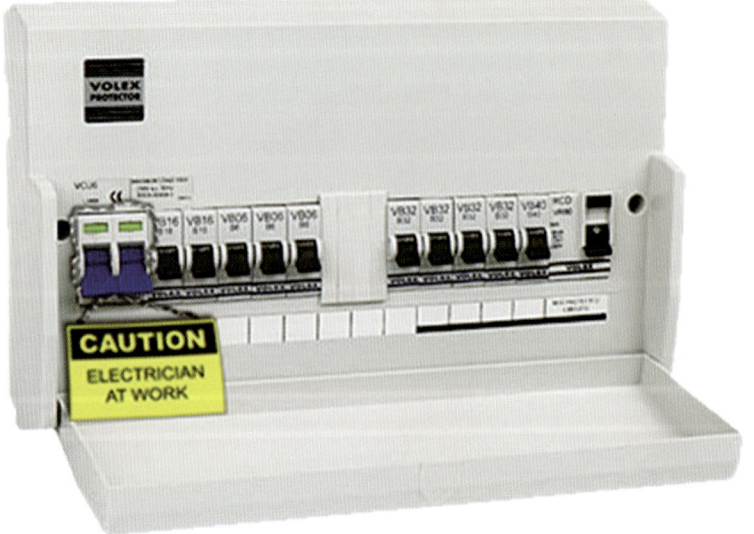

Step 6: Begin work

E-LEARNING

Use the e-learning programme to see a demonstration of safe isolation.

Energizing electrical installations

A number of deaths and major injuries have occurred when electrical circuits have been energized at the request of building designers, clients, contractors or finishing trades before the electrical installation was complete. It is not considered 'reasonable to work live' solely on the grounds of inconvenience, lost time or cost.

Electrical contractors are only able to energize circuits when it is unreasonable to work dead, and a written request has been made by the main contractor, or his agent. Suitable precautions and testing must also be undertaken before the electrical contractor agrees it is safe to energize the circuit.

Other aspects of electrical site safety

Here are some more ways in which you can ensure site safety with regard to electricity.

Using reduced voltage equipment reduces the risk of injury and the supply voltage should be limited to the lowest needed to get the job done. Battery-operated power tools are the safest.

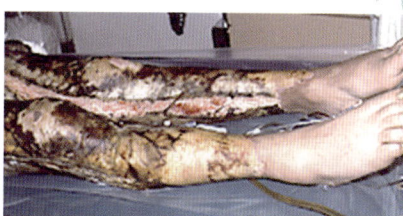
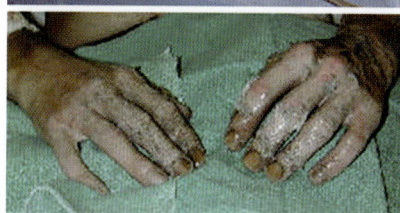

Many major injuries have occured when electrical circuits have been energized before installation was complete

Use battery-operated power tools – they are safest to use

Portable tools designed to run from 110V centre-tapped-to-earth supply are readily available.

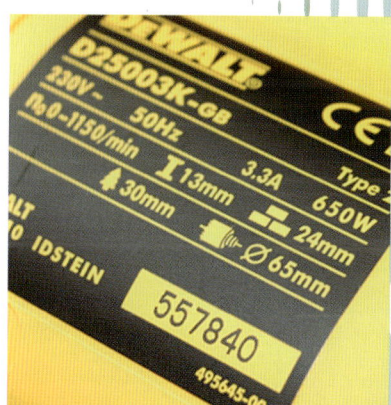

Use reduced voltage equipment to reduce risk of injury

Portable tools are readily available

Using a Residual Current Device (RCD) with 230V + equipment can reduce injury.

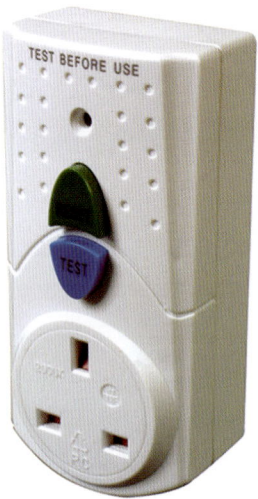

Using a Residual Current Device (RCD) can reduce injury

When working near overhead power lines these should be switched off if at all possible.

If work is near overhead power lines they should be switched off

CHECK YOUR KNOWLEDGE

1. **Imagine you have returned to a job with only a few tasks remaining. You have been wearing overalls and a hard hat, but your gear is in the van. The risks don't seem serious and the work will only take a few minutes. Do you have to use your PPE?**

   ☐ a. No – the work will only take a few minutes

   ☐ b. No – the risks do not seem serious

   ☐ c. Yes – the PPE should still be used

2. Under the Control of Substances Hazardous to Health Regulations – and European law – who has overall responsibility for controlling exposure to hazardous substances in the workplace?

☐ a. Employers

☐ b. Employees

☐ c. Government Health and Safety Inspectors

3. What items do you think should be part of a first aid kit? List seven items in the table below.

Item

4. Here are the five steps in the risk assessment process. Put the steps in the right order in the table shown.

☐ a. Who might be harmed and how?

☐ b. Review risk assessment and update if necessary

☐ c. Record findings and implement them

☐ d. Identify the hazards

☐ e. Evaluate the risks and decide on precautions

Step	Description
1.	
2.	
3.	
4.	
5.	

5. Which one of the following fire extinguishers is suitable for use on all classes of fire?

☐ a. Carbon Dioxide

☐ b. Foam

☐ c. Powder

☐ d. Water

6. If you were visually inspecting this piece of equipment how many faults would you find?

☐ a. One

☐ b. Two

☐ c. Three

☐ d. Four

7. Ladders and scaffolding are the means most installers use to enable them to work at heights. However, they are not always used safely. What percentage of accidents reported to the Health and Safety Executive do you think are due to falls by people working at heights?

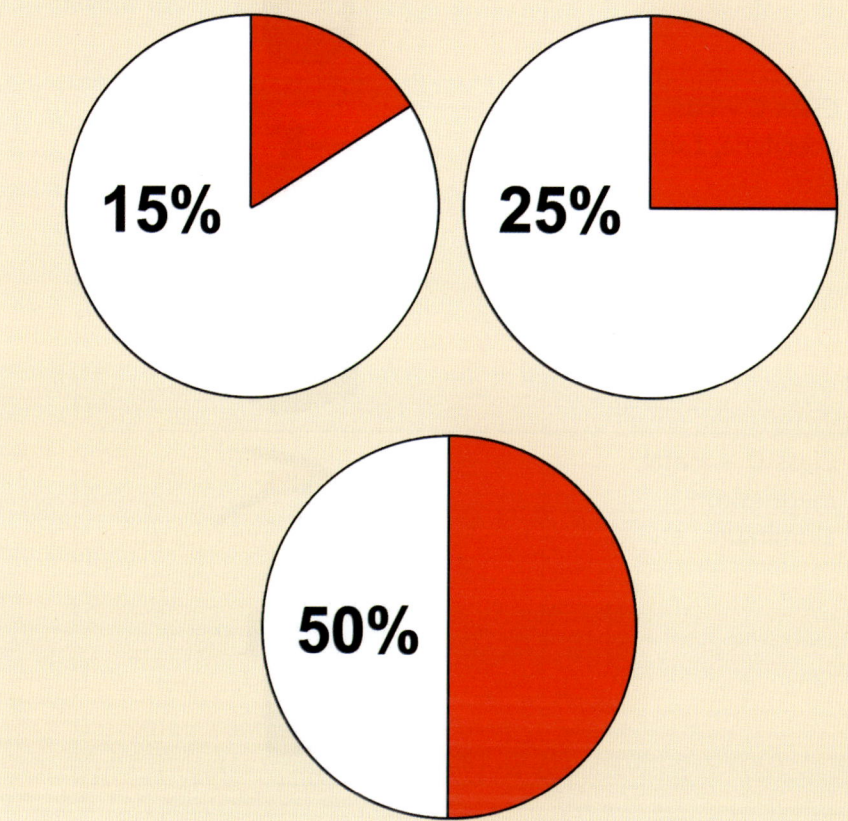

8. Match up the fire extinguishers shown with the class of fire they are used on.

Class of Fire	Fire Extinguisher
Class A – wood, paper, textiles, etc	

Class B – oil, petrol, paint, etc	
Class C – gas, acetylene, butane, etc	
Class D – metal, magnesium, aluminium, etc	

9. **In what order should the principles for controlling risks be applied?**
   Complete the table below by using the options below:

☐ a. Issue personal protective equipment

☐ b. Organize work to reduce exposure to the hazard

☐ c. Prevent access to the hazard

☐ d. Provide welfare facilities

☐ e. Try a less risky option

1.	
2.	
3.	
4.	
5.	

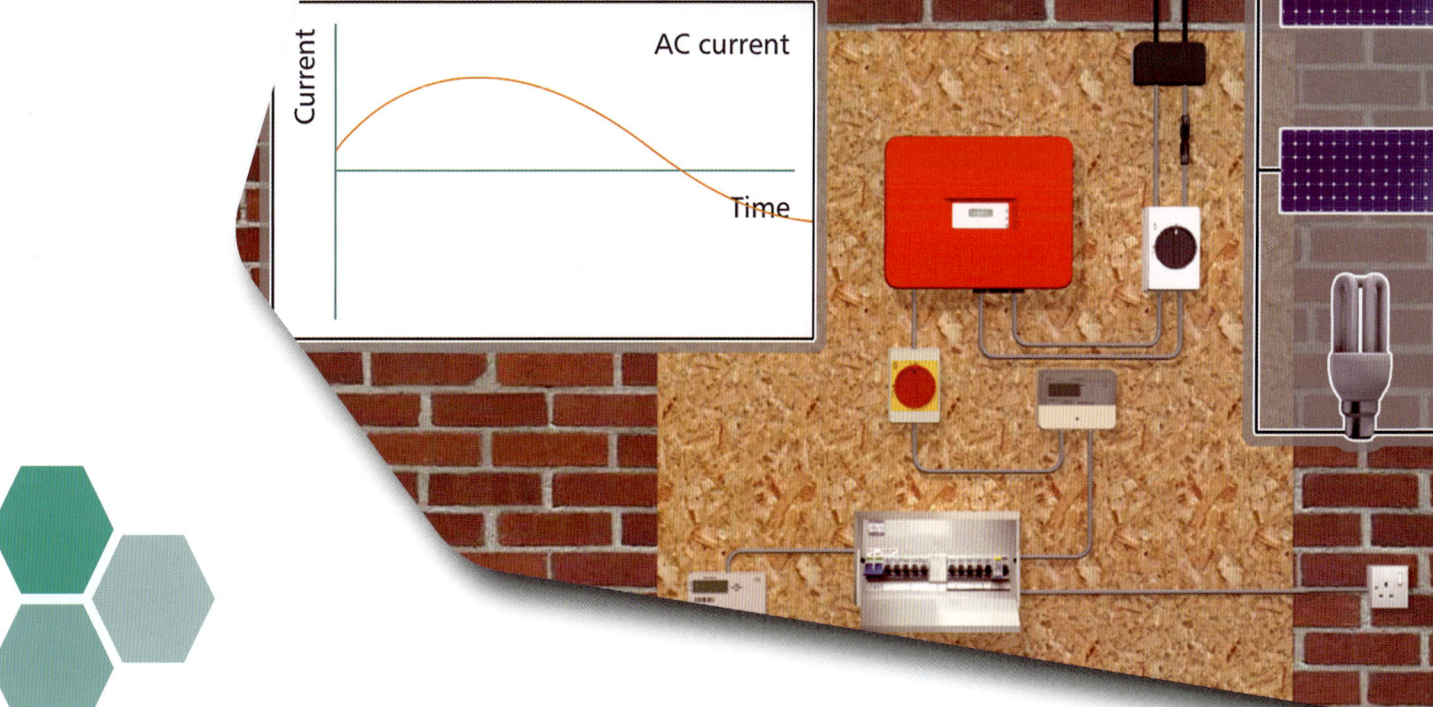

Chapter 3

SOLAR PV BASICS

LEARNING OBJECTIVES

By the end of this chapter you will be able to:

- Explain basic electrical principles used in solar PV systems

- Describe the importance of electrical isolation within solar PV systems

- Describe the main differences between grid connected and stand alone solar PV systems

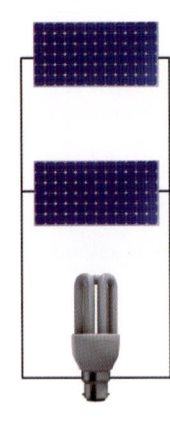

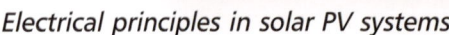

Electrical principles in solar PV systems

Solar PV on a motor home

AC AND DC CURRENT

Direct current (DC)
Direct current has a continuous voltage, and electrons flow in one direction. Solar PV cells and batteries produce direct current

A **direct current** is an electricity supply with a continuous current and it usually is simply referred to as DC current. Solar PV cells produce direct current and so do batteries. The electron flow making the current can be thought of as flowing in one direction only.

Graph showing direct current

Alternating current (AC) An electricity supply that has an alternating voltage, which changes the current direction many times a second, measured in Hertz (eg 50 Hertz)

Alternating current, or AC current, is an electricity current which continuously reverses direction. The electrons can be thought of as flowing one way, then the other, changing direction frequently. The change in direction happens many times a second, about 50 times and is measured in Hertz. Solar PV cells and batteries produce DC current.

Electricity suppliers use AC current because they can use step up and step down transformers to shift electric supply over large distances. Transformers cannot be used with DC current, which loses its power when sent over long distances.

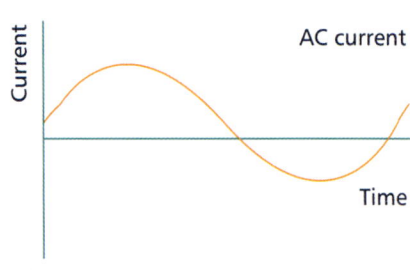

Alternating Current

Solar PV systems generally convert the generated DC current produced to an AC current, using an inverter. Inverters are easily integrated into domestic solar PV installations.

Solar PV systems convert DC current to AC current

SERIES AND PARALLEL CIRCUITS

The way an electric circuit is connected makes a difference to the current and voltage that may appear.

If the power source such as a PV module or **battery** is connected in a series, one after the other, then the voltage will build. If they are connected in parallel, then the current will build. The power output of each is the same.

Which type of circuit is required depends on what is being used in the circuit and whether the demand is for a higher current or higher voltage. We have a light bulb in the diagram, but it could be that the PV modules are being used to charge 12 volt batteries, or run a DC appliance.

Battery Used as a back up system in remote areas as an alternative to the grid or generator. Deep discharge batteries are used in PV systems. Batteries produce DC current

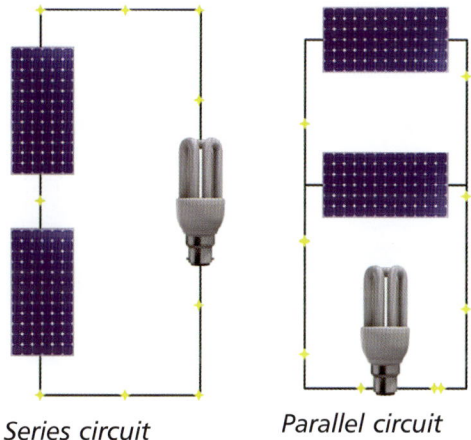

Series circuit   *Parallel circuit*

Series circuits

The series circuit has the power source connected one after the other. The voltages will be added together in this case giving a combined voltage across the light bulb of 30 volts. The current, however, remains the same as the current flows in a simple continuous circuit.

A series circuit

FUNCTIONAL SKILLS

15 volts + 15 volts = 30 volts
Current is 5 **amps** throughout the circuit.

Amps A unit of electrical current

Parallel circuits

A parallel circuit has the power sources next to each other with no direct connection between them, as the current flows in one direction only. This allows the current to be added together but the voltage across the light bulb will remain at 15 volts.

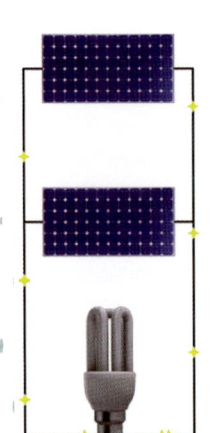

A parallel circuit

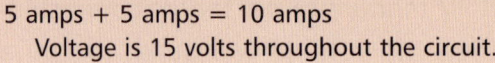

FUNCTIONAL SKILLS

5 amps + 5 amps = 10 amps
    Voltage is 15 volts throughout the circuit.

ACTIVITY 5

STC stands for standard test conditions and amongst other features, it includes 1000 kWh/m². A module has 48 cells with each cell measuring around 0.1m by 0.1m. Each cell in the module produces a nominal 0.5 volts and 1 amp of current at STC conditions. What is the theoretical maximum nominal voltage and theoretical maximum nominal current? How might you wire this number of cells (series or parallel) so that the cell could charge a 12 volt battery? What would be the difference between your PV panel design and a real world 12 volt PV panel?

PV UNITS OF MEASUREMENT

Volts, amps and watts

Volts measure the potential difference between two points – such as found between the positive and negative terminals on a battery, the two faces of a solar PV cell, or the supply to domestic electric sockets. Voltage has the symbol 'V'.

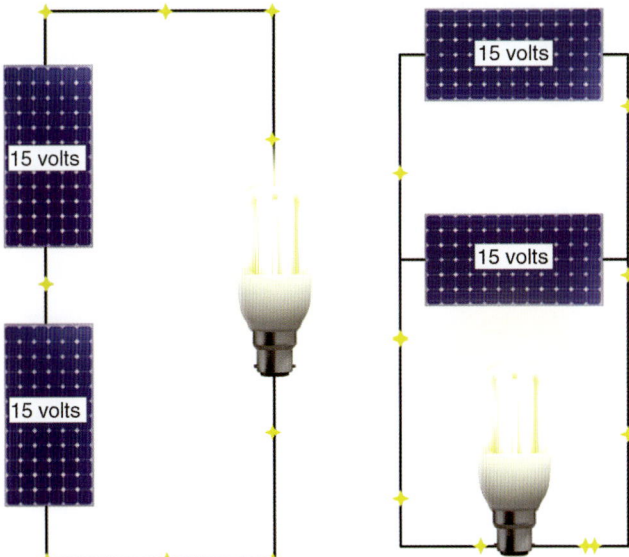

The series circuit has a voltage of 30 volts, and the parallel circuit has a voltage of 15 volts

The current is the flow of electrons in the circuit, measured in amps and shown in calculations by the symbol 'i'.

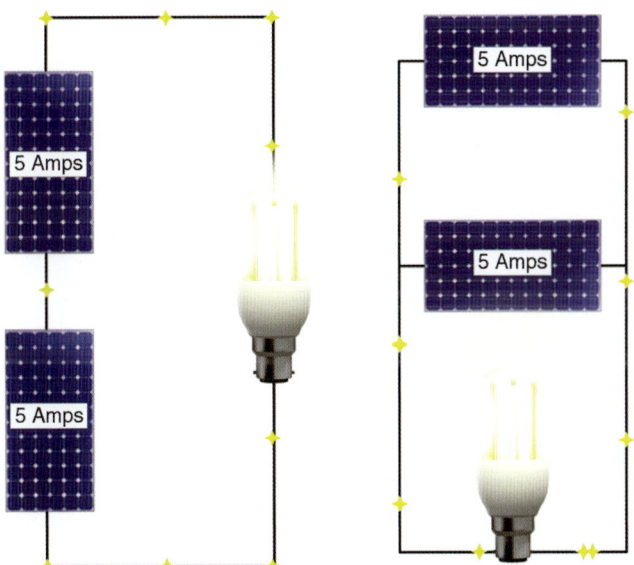

The series circuit has an amperage of 5 amps, and the parallel circuit has an amperage of 10 amps

Power is measured in **Watts** and Kilowatts, symbol W and kW. It indicates the rate of energy conversion from one form to another, or from one place to another. It is often measured in terms of time, such as hours which gives the units of energy used, shown as Watt hours (Wh). For example if a 1 kilowatt electric fire runs for one hour, it will have used 1 **kilowatt hour** of energy.

Watts Watts measure electrical power, and when linked to time give units of energy, shown as Watt Hours (Wh). Electrical power is calculated by multiplying current by voltage

Kilowatt hour (kW/h) Unit of energy, usually electrical, equivalent to a device that consumes or generates electrical power at the rate of 1kW for one hour. 1kWh is equal to 3 600 000 joules

Power can be calculated by multiplying voltage by current. In our examples the series circuit and parallel circuit both produce the same amount of power.

FUNCTIONAL SKILLS

Power is calculated by voltage times current, or Power = V × I

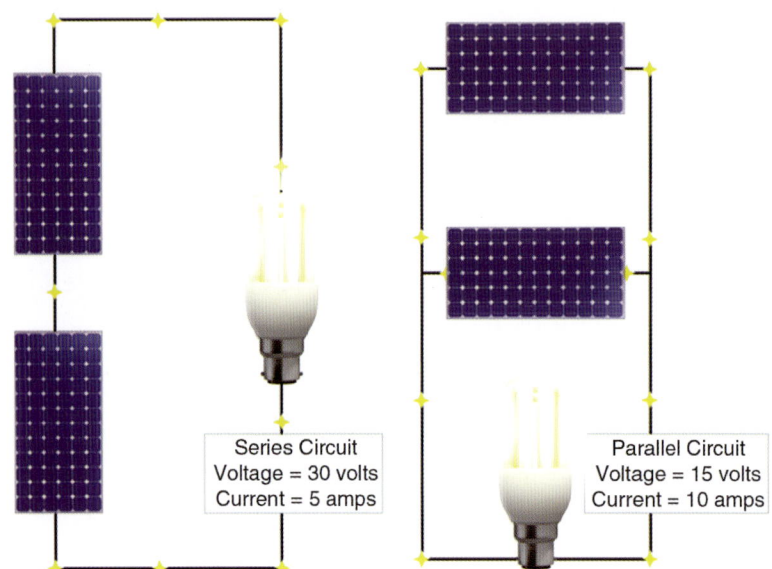

Series Circuit
Voltage = 30 volts
Current = 5 amps

Parallel Circuit
Voltage = 15 volts
Current = 10 amps

The series circuit has 30V × 5A = 150W of power and the parallel circuit has 15V × 10A = 150W of power

CONVERSION EFFICIENCY

The performance of a solar PV cell is measured by its efficiency for turning sunlight into electricity. Some sunlight is reflected, some passes through and is lost as heat, and some is absorbed, so a typical PV cell efficiency is about 14 per cent to 15 per cent. The **conversion efficiency** varies on the material used in the PV cell. Some materials have efficiencies of about 30 per cent, but these tend to be rarer materials and therefore more expensive. Low efficiencies mean larger arrays are needed. However, as the fuel is free, this doesn't usually matter. The PV industry is always working at developing more efficient PV cells.

Conversion efficiency of a PV cell A measure of the efficiency with which the PV cell, panel or array can convert the available solar energy into electricity. Most materials are about 14 per cent, but some rare materials have conversion efficiency of 30 per cent

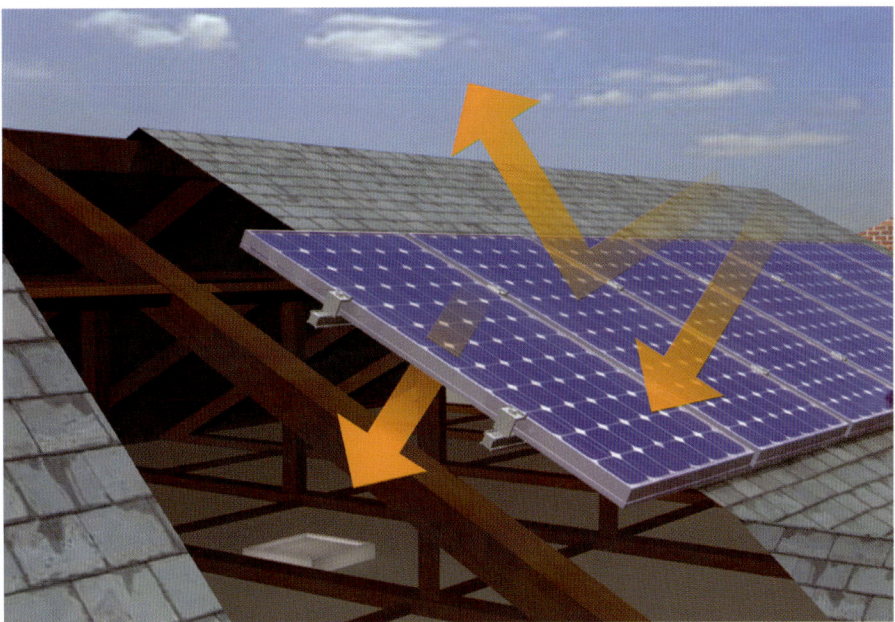

Some sunlight is reflected, some passes through as heat and some absorbed

Energy available for conversion

The Sun provides us with about 1kW/m^2 of power at the equator. This decreases with latitude but can still regularly reach 700W/m^2 or more in the British summer.

In total the Sun provides 8 300 times more energy onto the surface of the Earth than we use each year.

Sun provides 8 300 times more energy than we use each year

POWER OUTPUTS

Power outputs of PV cells

Information given on PV cells about their power output is always in terms of the maximum possible output. This is calculated for the brightest part of the day. Given the variability of the weather, especially cloud cover, the actual power output of a PV cell varies considerably over remarkably short lengths of time.

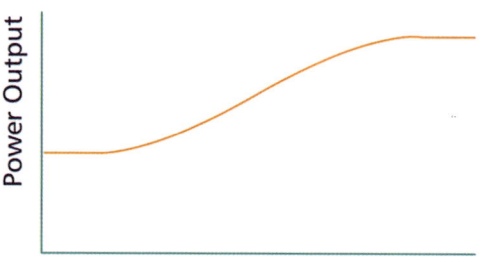

Graph illustrating how power output of a PV cell varies over time

Cell outputs vary considerably depending on the materials they are made from. A silicon cell will generate approximately 0.5V, so when joined together in series in a panel of approximately 1m × 2m the output can be up to 48V.

Generally, 20m² of PV panel is enough to supply about 50 per cent of a household's electricity requirements.

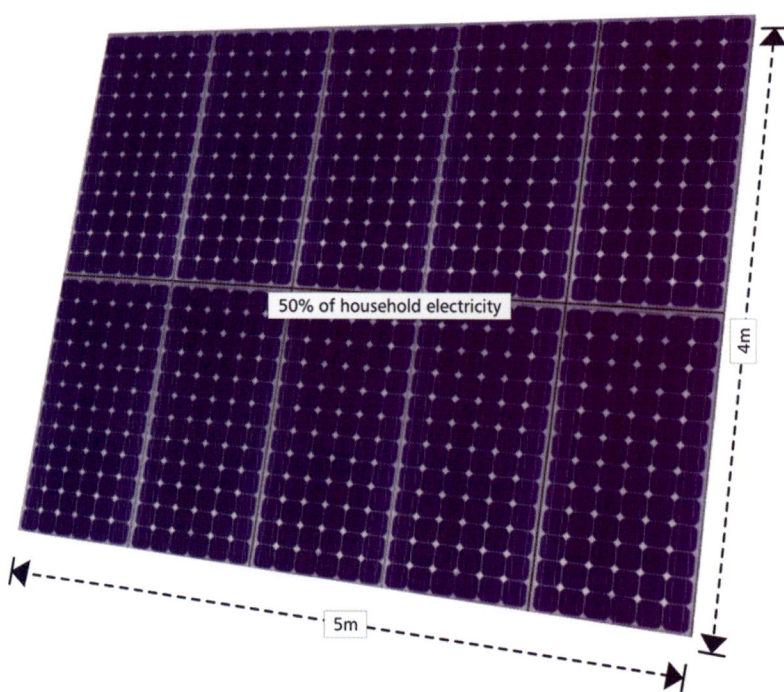

20m² of PV panel can supply 50 per cent of a household's electricity requirements

ACTIVITY 6

When quoting power output figures of a PV panel, the power output is quoted at STC conditions. Use the Internet to look up STC conditions and briefly explain what they are.

STAND ALONE AND GRID CONNECTED SYSTEMS

Stand alone systems

It is possible to power a home in the UK and other northern European climates independent of other sources of electricity using a stand alone system.

Worldwide, stand alone solar PV installations are probably more common than grid connected, due to developing world applications.

Stand alone systems can be used to power motor homes

If a stand alone system is the best option for a particular situation – such as a remote holiday home, it is important to design the system so the supply meets the expected demand **load**. The back up system provided by batteries would also need to be carefully sized to meet demand.

Stand alone systems can run entirely on DC current, or on AC current, or have some items, such as lighting, on DC current and the rest on AC current. An inverter would be used. Appliances are made

Load Anything in an electrical circuit that draws power from that circuit when turned on

specifically for use with either DC current or AC current – the appliances are not transferable! It is essential to use different plugs and sockets for DC appliances to prevent accidents.

Some appliances, such as fridges and TVs, are available for DC circuits, and many smaller appliances such as computers and mobile phones, already run on DC current.

Interior electrical workings of a remote holiday home

ACTIVITY 7

It says above that it is possible to power a home in the UK with just PV generated energy, i.e. a grid disconnected PV system. However, whilst this is technically feasible, in practice it isn't easy. What would be required to make this possible and what actions could be taken to simplify or reduce the cost of the system whilst keeping the system renewable? Would it be easier to design a system to run on just solar PV in Greece?

In developing countries the electricity grid tends not to reach rural areas. In this case DC current can be the best and least expensive way of providing electricity to these areas. Mostly it is used for lighting not only in the home but also in schools, hospitals and shops, prolonging opportunities for productivity in business and study. Solar PV power is also very important for supplying drinking water for irrigation. Other uses include refrigeration, TV and radio, electric fences for livestock control and generally improving living standards.

Stand alone systems are ideal for use in developing countries

Grid connected systems

SUSTAINABILITY

A mains connection means that the grid can provide additional electricity when there is insufficient sunlight to meet the electricity demand of the user. It also works in reverse in that excess electricity can be passed back via the grid to power other local houses and so power stations burn less fossil fuel. In the UK, grid connected PV and other renewable electricity sources gain both a generation tariff for all the electricity produced and an export tariff for all the electricity that is sold to the grid.

The grid connected system is a very sustainable solution

Inverter

A solar PV system produces DC current, and mains electricity is AC current, therefore an inverter which can convert DC to AC is essential for grid connected systems. Inverters convert the DC current to an AC current ready for export into the local distribution network, at about 230V. The local Distribution Network Operator will ensure the inverter is of the correct approved type. **Isolators** will also be added to the installation to cut the link to the grid should it be necessary.

Isolator A disconnect switch that can be used to provide safe isolation of an electric circuit enabling it to be worked on by an electrician

Inverter that can convert DC to AC is essential for a grid connected system

Meters

Grid connected systems need to have meters registering the production of electricity from the PV panels. Meters are also essential to record the electricity imported from the grid and generated by the PV panels. Existing meters for normal use as a domestic customer will only record imported electricity, so a new **generation meter** is required as part of the PV installation.

Generation meter Meter that records all the electricity generated by the PV array before any has been used by the household, or distributed to the grid

Meters are essential to register the production of electricity

E-LEARNING

Use the e-learning programme to learn more about grid connected systems.

ACTIVITY 8

Can a cheap simple DIY type meter be used to measure the energy production of the PV system? After all, the householder is not paying to purchase the electricity so why do they need to use an accurate high grade meter?

Grid connected options

In places where power cuts are frequent it is also possible to include batteries as an alternative **back up system**. This is also called an un-interruptable power supply (UPS).

> **Back up system** An additional source of electricity to ensure an electricity supply when the PV array is not producing sufficient power. Back up can be supplied by the grid, batteries or a generator

Batteries are a good alternative back up system

SAFE ISOLATION

HEALTH AND SAFETY

In all electrical systems it is important to be able to isolate the source of power from the rest of the circuit for maintenance purposes – and more importantly, to protect anyone who would be working on the system. Solar PV systems are not an exception from this rule. Even low DC voltages can be dangerous to work on.

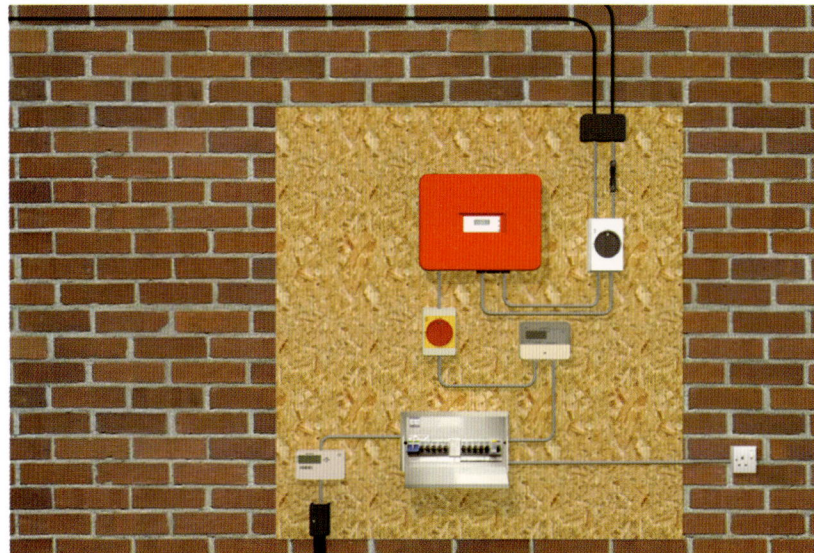

It is important to isolate the source of power for maintenance purposes

Stand alone systems

Stand alone systems need some way of isolating the batteries and other important equipment from the PV modules for maintenance and repair. Depending on the size of the array and connection in series or parallel, a sudden change in level of sunshine falling on the panel can make a large jump in the current produced.

Cross section of a remote holiday home

Grid connected systems

As for stand alone systems, there needs to be a way of isolating the PV system in grid connected circuits. Therefore, PV systems connected to the grid have a double pole isolator fitted between the generation meter and energy company meter. Also systems connected to the grid must have an automatic isolator feature to prevent feeding electricity to the grid in the rare circumstances that the grid is not operational – such as after a storm.

HEALTH AND SAFETY

Repairs may be carried out some distance from the domestic PV system and technicians repairing the grid must know that it is safe to work. This automatic isolation feature is built into the electrical inverter which looks for the AC current wave form. If it disappears, as in a power cut, the isolator cuts the grid connection and so makes the electrical circuit safe.

PV systems connected to the grid have a double pole isolator

ACTIVITY 9

Automatic isolation in the event of a power cut from the mains grid is an essential feature of grid connected inverters. What other features to minimize maintenance requirements and other benefits could be provided from a grid connected inverter?

CHECK YOUR KNOWLEDGE

1. **Answer the following questions about the image shown.**

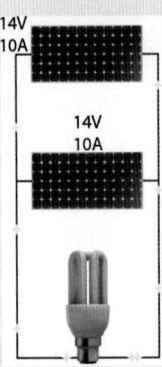

☐ a. What type of arrangement are the PV modules placed in?

☐ b. What is the voltage in the circuit?

☐ c. What is the current in the circuit?

☐ d. What is the power generated by the panel in full sunshine?

2. **Answer the following questions about the image shown.**

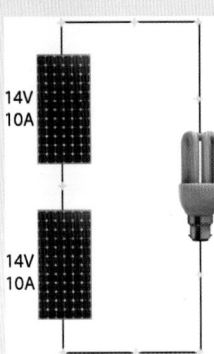

☐ a. What type of arrangement are the PV modules placed in?

☐ b. What is the voltage in the circuit?

☐ c. What is the current in the circuit?

☐ d. What is the power generated by the panel in full sunshine?

3. **What item of equipment, in a grid connected domestic PV installation, prevents electrical accidents to people working on an isolated local grid or during a power cut?**

☐ a. Inverter

☐ b. Transformer

☐ c. Meter

☐ d. Connectors

Chapter 4

TYPES OF PV MATERIALS

LEARNING OBJECTIVES

By the end of this chapter you will be able to:

- Describe the different types of solar PV materials available

- Describe the relative efficiencies of PV materials

- List the materials from which PV modules may be constructed

- Describe at least three different applications for solar PV materials

Type of photovoltaic materials

TYPES OF PV MATERIALS

There is a short list of materials sufficiently sensitive to sunlight to capture the solar energy and turn it into electricity. Fortunately the most important is silicon, which is used to make over 95 per cent of all solar PV cells, it is also the second most abundant element found on Earth, mostly as sand. Other materials sometimes used are more efficient but also more expensive and much rarer, making their use more limited. A great deal of research and development is currently being done to improve the efficiency of currently used materials and find other options.

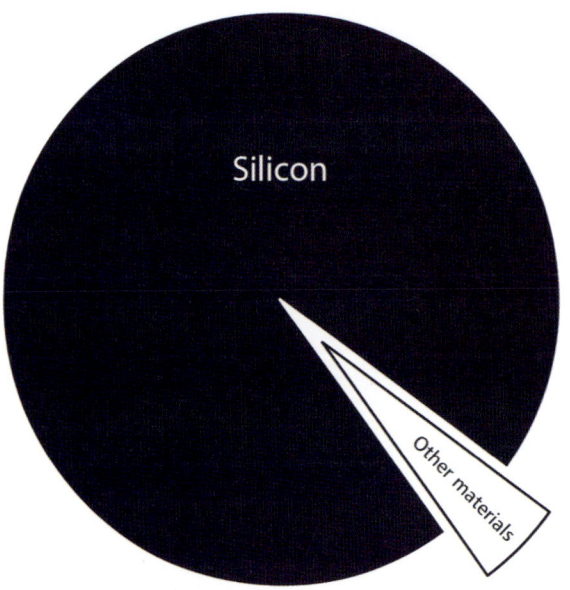

95% of all solar PV cells are made of silicon

ACTIVITY 10

Silicon is also widely used in another electronic component that has a significant influence on modern society. Name this component and state how it is different to a piece of silicon PV.

Monocrystalline silicon cells

Each PV cell of **monocrystalline silicon** is made from a wafer of thin crystal. The silicon is cut from cylinder-shaped ingots and so does not completely cover a square PV cell module without wasting silicon. This leaves uncovered gaps at the four corners of the cells. This makes them easy to recognize, giving a distinct pattern.

Monocrystalline silicon cells are the most expensive, however they are the most efficient silicon option, therefore the smallest panels are required to produce the same electrical output.

Monocrystalline silicon Silicon solar cells cut from a single silicon crystal. These are the most efficient form of silicon for PV use

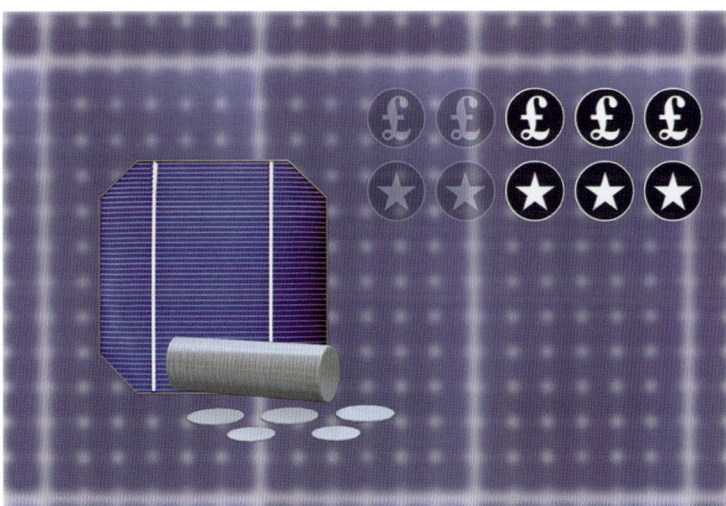

Monocrystalline PV cells

Polycrystalline silicon cells

Polycrystalline silicon cells have a 'marble-like' appearance with many small silicon crystals making the PV layer. It has a sparkly blue appearance. It can be made in square ingots and then cut into thin wafers in the same way as monocrystalline PV cells. By making the small crystals slightly larger and aligning them so the light penetrates deeper, it is possible to improve the efficiency of the cell; however

Polycrystalline silicon Crystalline silicon where the crystals are positioned in a random pattern. Polycrystalline silicon is slightly less efficient PV material than monocrystalline silicon but much easier to make

polycrystalline PV materials are slightly less efficient than monocrystalline. Polycrystalline silicon cells can also be made into ribbons from molten silicon rather than square cells.

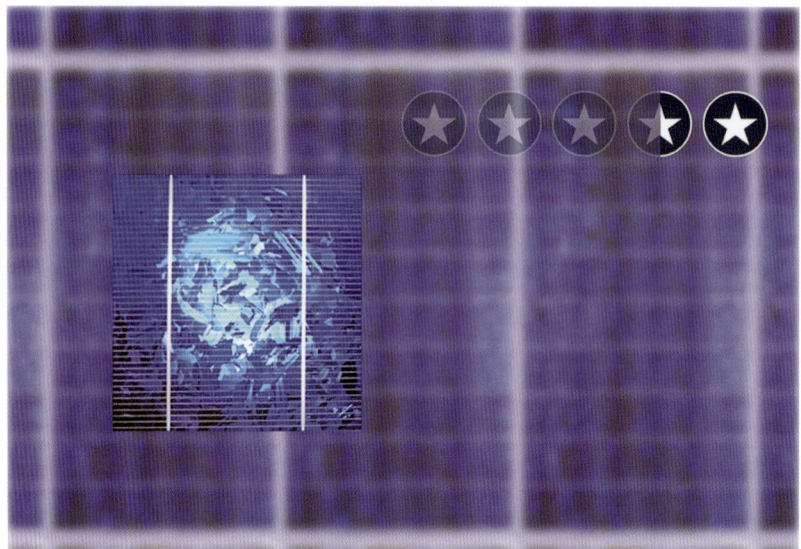

Polycrystalline PV cells

Thin film technology (also called amorphous)

Thin film, or amorphous PV panels are manufactured by coating a layer or layers of silicon onto a substrate. Because thin film only requires a shallow layer of silicon, it reduces the amount of light absorbing material required to create a PV cell. This makes them the cheapest to produce but also results in the lowest efficiency. Unfortunately the material does degrade over time and therefore loses some of its efficiency. Amorphous PV materials may be made of silicon or other light sensitive substances.

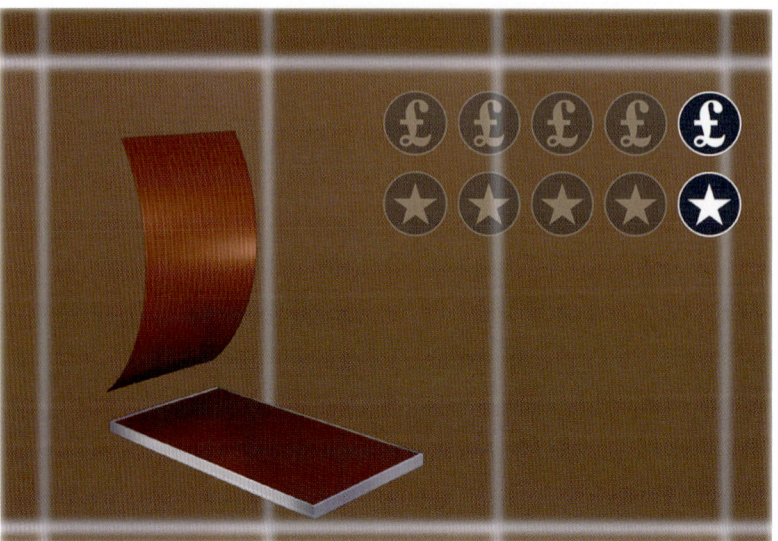

Thin film technology

Second generation

Thin film technology allows lighter and more flexible structures to be covered in the PV material. Therefore, it can be coated onto a variety of substrates such as back-packs.

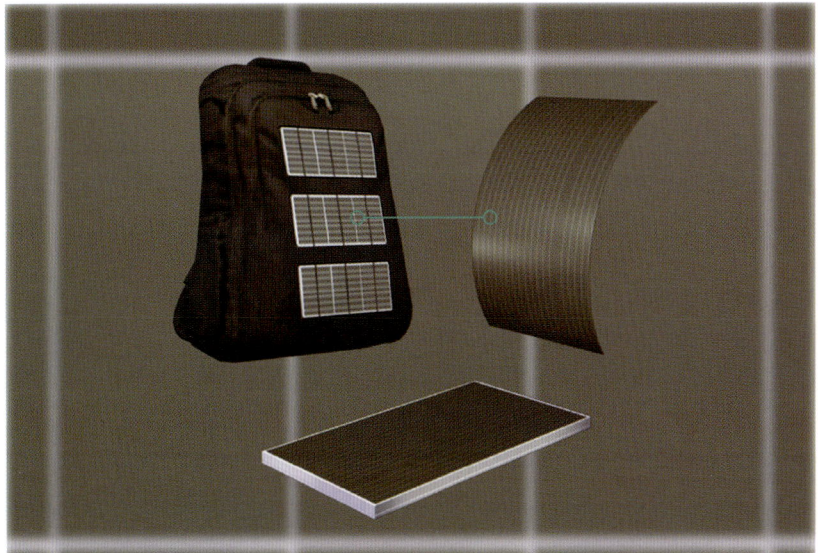

Second generation PV materials

Third generation

Third generation PV is not yet with us but research is being carried out using nanotechnology which may lead to very thin stacked plastic layers with high efficiency and low cost.

Third generation PV materials New materials developed through nano-technology for increased efficiency and lower cost

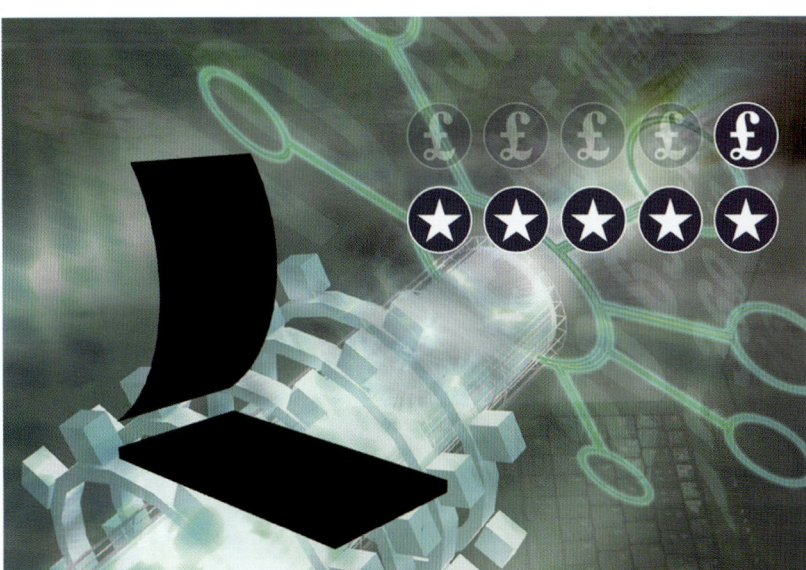

Third generation PV materials

Hybrid cells

Hybrid cells Cells made from a mix of crystalline and amorphous materials in layers tuned to different wavelengths of light

Triple junction PV cells An alternative name for hybrid cells

Tandem junction PV cells An alternative name for hybrid cells

Hybrid cells are a mix of amorphous and crystalline PV materials organized to make maximum use of the Sun's radiation. Silicon can be 'tuned' to pick up different wavelengths of light. By having layers sensitive to several distinct wavelengths, stacked construction of the PV cell can improve the efficiency as each layer extracts energy from its specific wavelength.

Hybrid cells sometimes offer the best value for money based on electricity savings against capital cost, although this does vary on the current silicon price. Hybrid cells may also be called **triple junction** or **tandem junction** cells.

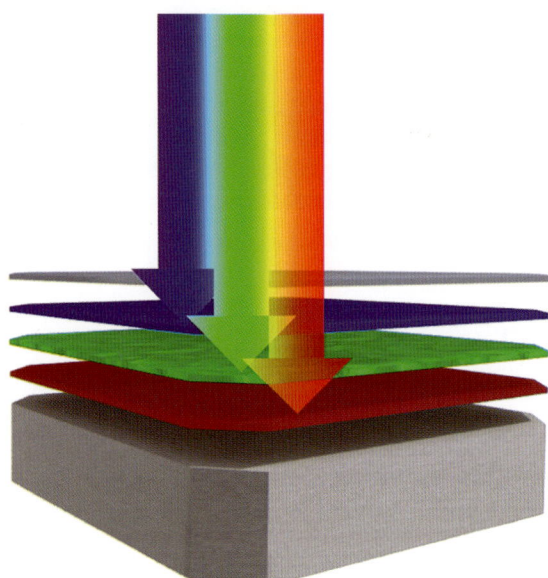

Hybrid cells - mixture of amorphous and crystalline PV materials

ACTIVITY 11

Thinking about monocrystalline, polycrystalline, thin film and hybrid collectors, please briefly name some advantages (and/or disadvantages) for each type of material.

Non-silicon based PV materials

Compounds other than silicon are also used as PV materials. They are generally more efficient than silicon but also more expensive in either cost of material or production. Unlike **amorphous silicon**, amorphous PV modules made from other materials do not degrade over time. The main compounds used are:

- **Gallium arsenide** (GaAs)
- **Cadmium telluride** (CdTe)
- **Copper indium gallium selenide (CIGS)**

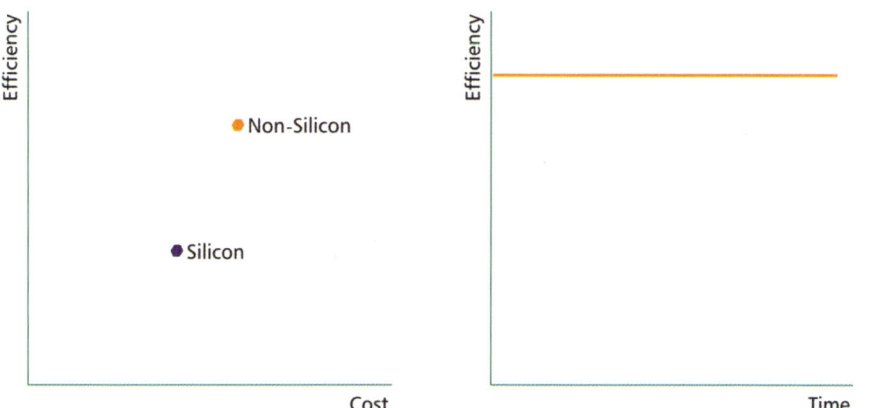

Non-silicon and more expensive & efficient; Silicon - less expensive and efficient

Gallium arsenide

Gallium compounds, of which gallium arsenide is the most commonly used, are mostly used as solar cells powering space stations and satellites. They have high efficiencies but also high production costs, so are used where the cost can be justified.

Gallium arsenide can be used in crystalline or amorphous form. It is capable of high temperature operation without loss of performance, and therefore is sometimes used where light concentration systems are selected, such as arrays in the desert. Unfortunately both gallium and arsenic are expensive scarce resources, and arsenic is toxic!

High efficiencies and production costs

Arsenic is toxic

Amorphous silicon
Silicon where the atoms are not arranged in a crystalline structure. Amorphous silicon can be deposited on glass and other structural materials in a very thin film using less silicon than crystalline technologies

Gallium arsenide A semiconductor used in PV cells, relatively more efficient than other substances but also rare and expensive

Cadmium telluride A semiconductor used in PV cells, highly toxic in production

Copper indium gallium selenide (CIGS) A semiconductor used in PV cells

Cadmium Telluride

Cadmium Telluride is used for thin film PV applications and a relatively cheap and simple electroplating type process can be used to make the modules. However cadmium is a highly toxic substance and many safety precautions and regulations are needed during production and disposal. Several manufacturers have withdrawn from production.

Relatively inexpensive

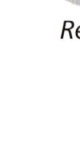

Highly toxic

Copper compounds

Copper compounds have been very promising in PV research and copper indium gallium senide or CIGS, has achieved the highest thin film efficiencies so far. Both copper-based PV materials use indium. Indium is both rare and expensive, but is also used to produce flat screen televisions and monitors. Seventy per cent of the world's indium production is currently used for monitors and therefore there may be some conflict if, or when, CIGS production is increased. Toxic hydrogen selenide gas is used during the manufacture of both CIS and CIGS products.

70% of the world's indium is used for flat screen television/monitor production

E-LEARNING

Use the e-learning to learn more about non-silicon based PV materials.

ACTIVITY 12

Many materials, including organic materials, have been shown to demonstrate the photovoltaic effect. What are the current main disadvantages of organic photovoltaic cells as compared to inorganic photovoltaic cells?

MANUFACTURE OF PV MATERIALS

Manufacture of crystalline silicon

Two types of crystalline silicon are used in PV materials. Monocrystalline was developed in the electronics industry and is very pure, with PV cells made of a thin wafer from a single silicon crystal. Polycrystalline PV cells are less 'pure' with many smaller crystals held within each cell. These small crystals are visible and give polycrystalline modules their sparkly appearance.

Crystalline silicon A form of silicon in which the molecules are ordered. If the crystals are orientated in one direction, it is known as monocrystalline form. If the crystals are more randomly organized it is a polycrystalline form. Monocrystalline silicon has a better conversion efficiency compared to polycrystalline silicon

Monocrystalline cells   *Polycrystalline cells*

Monocrystalline cells

Monocrystalline silicon PV cells are made from very pure molten silicon, crystallizing around a single seed to form a single solid crystal. The crystal is then cut into thin wafers with a distinct rounded off or squared shape. The system is slow and expensive.

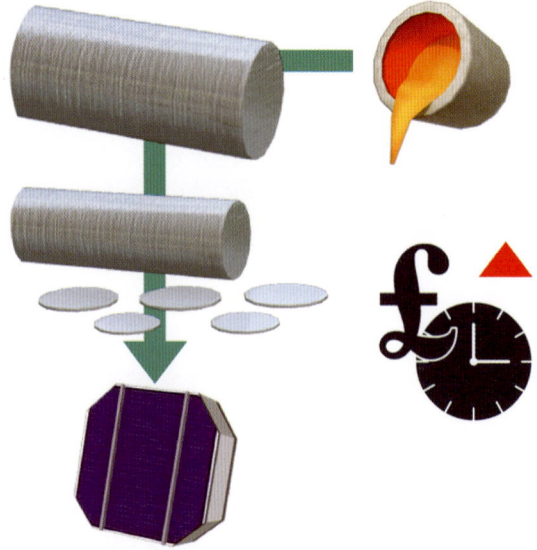

Monocrystalline cells

Polycrystalline cells

Polycrystalline silicon cells are also made from molten silicon, although the temperature is not as high and the production is quicker. Ribbons, tubes or ingots are 'pulled' from the silicon melt and the small crystals form randomly as it cools. By aligning the crystals during formation the resulting material is more efficient, although the aligning process adds to the manufacturing cost.

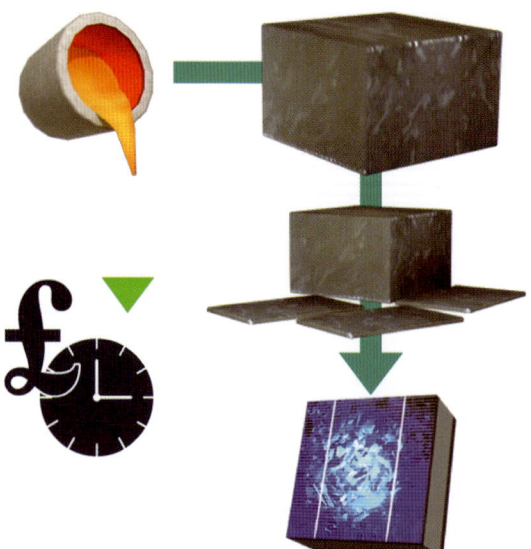

Polycrystalline cells

Manufacture of thin film PV materials

Amorphous or thin film PV materials are made by adding a very thin layer to a supportive material, such as glass, steel, slate or plastic.

Manufacturing is through a continuous low temperature process and is much cheaper than making crystalline forms of PV modules. Because the film can be applied to many types of substrate it lends itself to building **integrated applications**.

Integrated applications
Buildings where PV materials form part of the building materials, rather than panels added on a framework

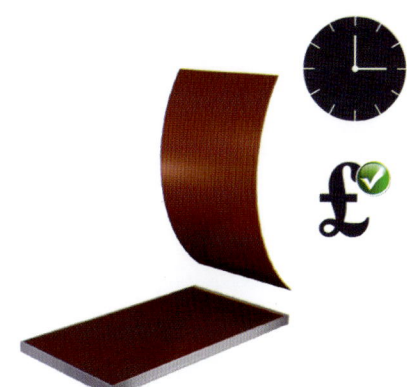

Thin film PV material

ACTIVITY 13

If you were asked to make some home-made solar panels, how might you go about this process? Clue, search the Internet for home-made solar panels.

CONVERSION EFFICIENCIES

Typical conversion efficiencies

The efficiency of each type of PV material varies, and needs to be balanced against the characteristics of a particular PV material when deciding which type of PV to use in a particular application. The efficiency rating is given at peak performance in optimum conditions.

PV Material	Efficiency	Advantages	Disadvantages
Monocrystalline Silicon	14-22%	High efficiency	Expensive to produce
Polycrystalline Silicon	10-14%	Cheaper to produce Easily shaped	Lower efficiency
Amorphous Silicon	6-8%	Good production on hazy days Cheapest to produce	Degrades Lower efficiency
Gallium Arsenide	Up to 39%	Tolerates high temperature	Scarce resources Very expensive
Cadmium Telluride	12-13%	Efficiency does not degrade Cheap manufacturing costs	Toxic in manufacture
Copper Indium Gallium Selenide (CIGS)	12-13%	Efficiency does not degrade	Indium scarce & expensive Toxic in manufacture

Table showing the efficiency, advantages and disadvantages of PV materials

Monocrystalline

Monocrystalline cells have relatively high efficiency and are widely used. Efficiency can reach above 22 per cent. However they are expensive to produce.

Monocrystalline cell

Polycrystalline

Polycrystalline silicon cells are cheaper to produce than monocrystalline and can be shaped for specific situations, however they are less efficient. Efficiency rates are between 10 per cent to 14 per cent in direct sunlight.

Polycrystalline cell

Amorphous silicon

Amorphous silicon has a much lower efficiency, and degrades over the decades, but gives good production on hazy days. The efficiency is only about 6 per cent to 8 per cent, however the energy source is free so can be compensated for by increasing the surface area of the array. The production costs are the cheapest.

Amorphous silicon

Gallium Arsenide

Gallium Arsenide has the highest efficiency rating of up to 39 per cent. It can also tolerate high temperatures, which other PV materials do not. The disadvantage is that the elements are rare and panels expensive to produce.

Gallium arsenide cell

Cadmium Telluride

Cadmium Telluride has a relatively good efficiency of around 12 per cent to 13 per cent and that efficiency does not degrade. It is also relatively cheap to produce. However cadmium is a highly toxic substance.

Cadmium telluride cell

CIGS

Copper Indium Galium Selenide – or CIGS – has a reasonable efficiency of around 12 per cent to 13 per cent, cheap manufacturing costs and does not degrade. Unfortunately indium is expensive and manufacture requires the use of toxic hydrogen selenide gas.

Copper indium gallium selenide (CIGS) cell

Implications of conversion efficiency

The implication of different efficiencies of PV materials affects the size of installations required to produce a set amount of annual wattage for a household. It needs to be considered when choosing solar panels.

The installation cost, including number of panels required, needs to include the structural effects of panels if they are to be roof-mounted. Also the amount of cabling required can also add to the cost.

It is often said that monocrystalline installations cost about 20 per cent to 30 per cent more than the equivalent polycrystalline installation and that polycrystalline is approximately 50 per cent more expensive than the equivalent amorphous installation. However, mass production is reducing these price differences and amounts relative to each other.

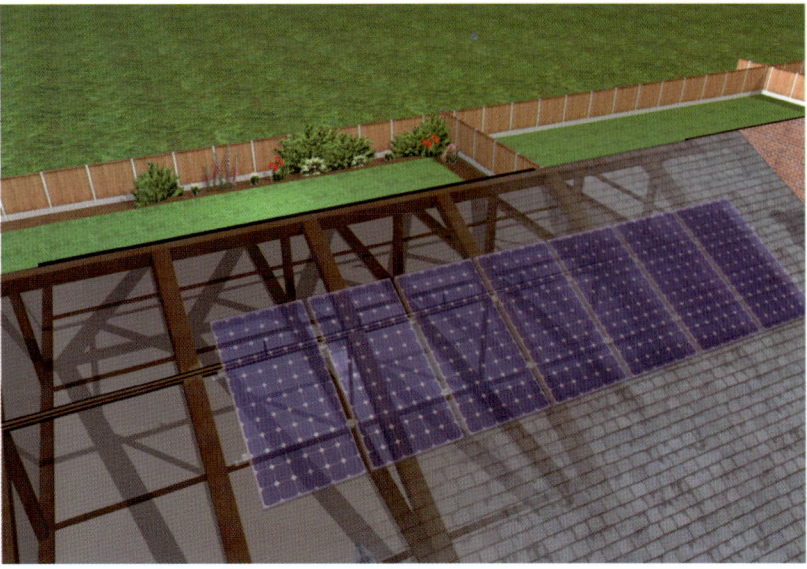

Installation costs include number of panels, and their structural effect.

Cross section of a roof top with solar panels installed

ACTIVITY 14

Which is the cheapest solar panel/kWh produced by the panel?

VARIETY OF PV APPLICATIONS

Small scale applications

PV cells have been used for small scale applications for some time and are particularly used where DC current is needed, such as radios, calculators, phone chargers, lighting of road signs and garden paths. Other applications include small refrigeration units and wells in developing countries. Flexible thin film technologies have been used in sails for small boats to provide electricity for long voyages. Electric cars can also be powered by solar PV and domestic cars with panels on their roofs are starting to be produced.

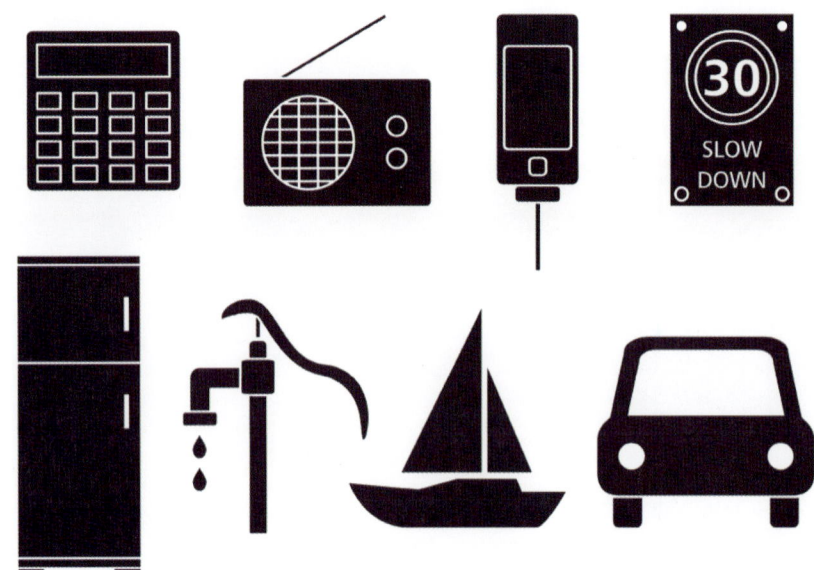

Examples of small scale applications where PV cells have been used.

ACTIVITY 15

Solar powered electric cars have been used to race across Australia. What influence are these solar racing cars likely to have on the mainstream car industry?

Domestic applications

Domestic PV systems can be added on to existing houses and flats although new builds have a wider choice of how to integrate the PV panels. If it is a new build, or the house is having a new roof, PV tiles rather than PV modules can be used. In England and Wales, existing roofs can have PV panels added, usually without planning permission, so long as they do not extend more than 200mm above the existing roofline. This also has to allow for **ventilation** under the panels to keep them cool and prevent overheating.

Flat roofs and garden-located PV can have panels installed at the optimum angle.

Caravans and holiday homes can be powered entirely by PV panels used with batteries.

Ventilation
Ventilation affects efficiency of PV panels, as they are less efficient in higher temperatures

PV tiles on new buildings

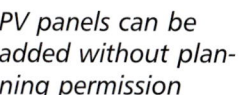

200mm gap needed when installing new plates

PV panels can be added without planning permission

Garden located PV panels can be installed at the optimum angle

Cross section of a motor home with PV panels installed

ACTIVITY 16

In the previous section it comments that in England and Wales, existing roofs can have PV panels added, usually without planning permission, so long as they do not extend more than 200mm above the existing roofline. Why has this 200mm been chosen and what might be important aesthetic property considerations when designing a solar PV system with a typical domestic property?

Large scale applications

Many industrial and commercial properties now have some of their power supplied by solar power. These can be added but often new buildings are now designed to maximize benefits of passive and active solar radiation. Companies also make use of low value land and install large arrays.

Power plants have also been built. Germany has a 340kW power plant at Kobern-Gondorf, others are found in USA, Italy and Switzerland. Power plants in areas nearer the equator with greater annual solar radiation are an attractive economic investment and can produce totals more than twice those in northern Europe.

Illustration of an example of a large scale application

Building integrated applications

Building integrated applications have part of the structure of the building made of PV materials. These are often thin film and amorphous materials rather than rigid crystalline panels, as their flexibility allows shapes to be fashioned to fit the building requirements and the variety of substrate allows the units to be strong enough as building materials. Building Integrated PhotoVoltaics is called BIPV and there are several specialist manufacturers working in this field, some of whom are building crystalline PV into windows and roofs.

Use of tiles avoids problems with ventilation and snow load issues, and cladding with PV materials can actually be cheaper than traditional materials.

Walls and windows can all be PV materials, allowing in light and reducing glare and at the same time producing electricity.

PV tiles are often made of thin film and amorphous materials - flexible to fit required shape

CHECK YOUR KNOWLEDGE

1. **For each of the different PV materials listed, select their approximate efficiency from the values listed.**

☐ a. Monocrystalline silicon – 0% / 5% / 10% / 15% / 20% / 25% / 30% / 35% / 40% / 45% / 50%

☐ b. Polycrystalline silicon - 0% / 5% / 10% / 15% / 20% / 25% / 30% / 35% / 40% / 45% / 50%

☐ c. Amorphous silicon - 0% / 5% / 10% / 15% / 20% / 25% / 30% / 35% / 40% / 45% / 50%

☐ d. Gallium arsenide - 0% / 5% / 10% / 15% / 20% / 25% / 30% / 35% / 40% / 45% / 50%

☐ e. CIGS - 0% / 5% / 10% / 15% / 20% / 25% / 30% / 35% / 40% / 45% / 50%

2. **Match the type of crystalline silicon cells with the correct image.**

Cell	Image
Monocrystalline cells	
Polycrystalline cells	
Amorphous silicon	
Gallium arsenide	
Cadmium telluride	
Copper indium gallium selenide (CIGS)	

3. **Which PV material has the highest efficiency for producing electricity from solar power?**

☐ a. Crystalline silicon

☐ b. Gallium arsenide

☐ c. Cadmium telluride

☐ d. Copper indium gallium selenide (CIGS)

Chapter 5

SYSTEM COMPONENTS

LEARNING OBJECTIVES

By the end of this chapter you will be able to:

- List the components found within solar PV systems

- Describe the function and purpose of each item

- Describe the components of a typical stand alone solar PV system

- Describe the components of a typical solar PV system exporting to the national grid

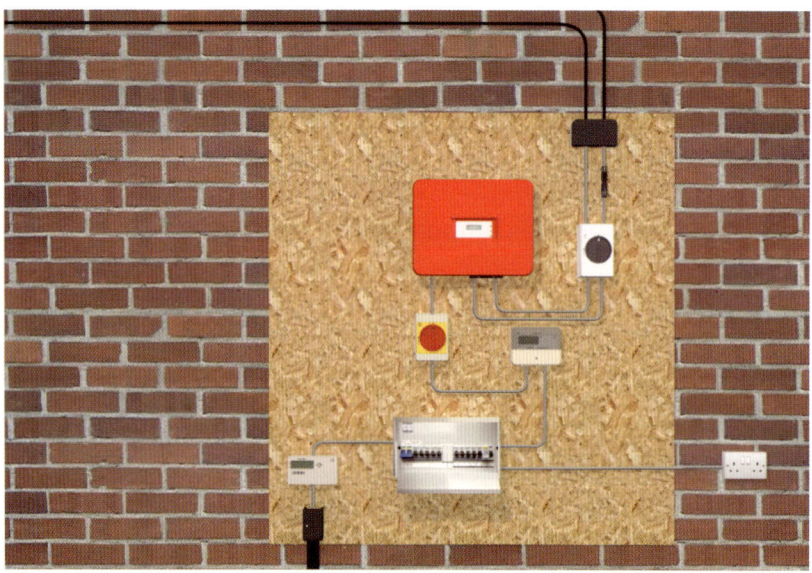

Components within a solar PV system

SYSTEM COMPONENTS

The basic components

There will always be a need to smooth out variations in electric supply and act as a back up supply after dark.

Stand alone systems may be chosen from necessity or choice, but will always require batteries to carry out this smoothing effect. For a grid connected system the smoothing is carried out by the grid as specified by regulations G83/1 and G59/1. Where the grid supply is erratic or unreliable it is OK to have batteries in addition to a grid connection.

Basic components of solar PV systems

Solar panels and mounting frames

Solar panels and their mountings are independent of the type of installation, whether grid connected or stand alone. How they are connected together (series or parallel) is important, dependent on the output requiring greater current or voltage.

Solar panels and their mounting frames

ACTIVITY 17

When mounting PV panels on a flat or pitched roof, the roof must be left in the same (or hopefully superior) condition as before the intervention. Discuss what is meant by 'in the same condition' and what this is pointing at.

Cabling and connectors

DC current requires heavy duty cabling. Therefore linking together the solar array, batteries and inverter, where DC current is carried, will require heavy duty cables and connectors. To maintain efficiency, these cables should be as short as possible. They must also be rated for the maximum **short circuit current**. AC current does not require such heavy duty cabling but the cables still need to be the correct size for the current.

Short circuit current (I_{SC}) The current flowing freely from an illuminated PV cell or module through an external circuit that has no resistance. The I_{SC} is the maximum current possible

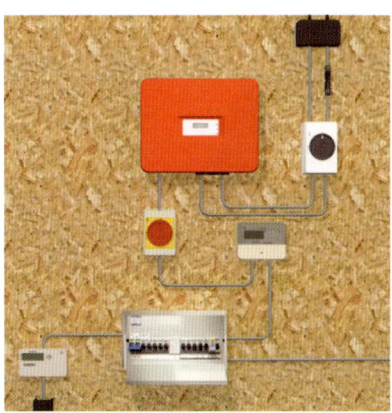

Heavy duty cabling is required for DC current

Isolators, switches and fuses

Isolators are usually made for either DC or AC current – they must not be used in the wrong circuit. Switches and fuses in the DC circuit must also be specified for DC current, as ordinary AC switches and fuses will

Isolator, switches and fuses

cause a hazard. It must be remembered that cables between the solar panels and the isolator will be live at all times during daylight unless the PV panels are covered with a thick black opaque covering.

ACTIVITY 18

What colour are AC and DC isolators, red or black?

Junction boxes and distribution boards

Up to two junction boxes and distribution boards may be required, one for each DC and/or AC circuits. If both AC and DC are used to supply lights and appliances, the inverter would be placed between the junction boxes.

Junction boxes and distribution boards

Batteries and controllers

Batteries are usually charged during the day and used at night. If solar production of electricity falls short during the day then battery power can provide the additional requirement. The controller manages the flow of electricity into and out of the battery.

ACTIVITY 19

What are the main differences between a car battery and a solar battery?

Batteries and controllers

Inverter

An inverter turns the DC current to AC current and also steps up the voltage. Because of losses in the inverter, it is more efficient to use the PV power as DC current; however other considerations need to be taken into account. If the system is to be linked to the grid, it must have an inverter as the grid cannot accept DC current directly.

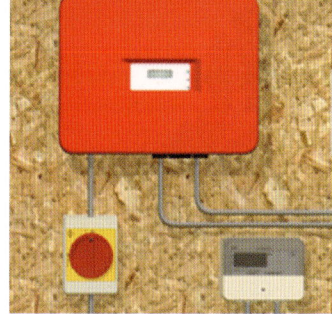

Inverter turns the DC current to AC current

ACTIVITY 20

What is the main difference between a grid connected and grid disconnected inverter?

Meters

Various meters are available, giving digital data on various usage and production parameters. Meters are particularly important for grid connected systems where excess electricity is fed back to the grid. PV systems will always require one or two new meters as a generation meter will have to be fitted and, frequently, the existing import meter might also be replaced.

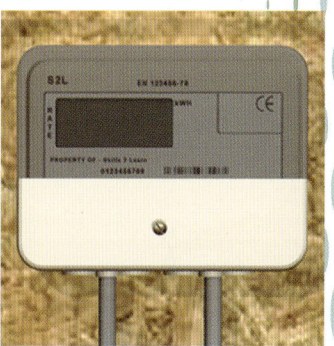

Meters

ACTIVITY 21

Meters record energy used. If you were going to build your own energy meter using two multimeters and a laptop, how might you go about recording the energy output of the system?

Labels

As with all electrical installations, all wires and connections should be clearly labelled as good practice. In all properties with solar PV all meters, junction boxes and isolators must be labelled to clearly state there are two sources of electrical supply to the system.

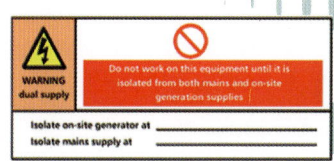

Connections should be clearly labelled

ACTIVITY 22

There is a Government sponsored guide available on the Internet (you can download it as a pdf) called 'Photovoltaics in Buildings, Guide to the installation of PV systems'. Look up in this guide the label that should be used on a DC cable and also the distance between labels.

Labels It is good practice to label cables and wiring. It is especially important to have PV systems labelled to show there are two sources of electrical supply

Safety issues

HEALTH AND SAFETY

There are several important safety issues with PV systems which are emphasized here and relate to keeping the AC and DC equipment separate. AC and DC cables must not be mixed or used together, or used on the other type of circuit.

Equipment used in low voltage circuits with less than 30 amp current can have standard mains switches and light sockets, BUT if standard plugs are used, the low voltage device could possibly be accidentally connected to a high voltage mains circuit – in which case the result would be disastrous! To avoid this happening non-standard plugs are used, such as those used in caravans and boats.

Just to be clear and set out what we mean twice, only use DC components in DC circuits and AC components in AC circuits. Always work to manufacturers' instructions and only use components rated for the relevant current, voltages and frequency (Hz) in the relevant section of the circuit. The Internet available booklets, "Guide to Photovoltaics", "Guide to the installation of PV systems" sets this and other specific guidance out in more detail.

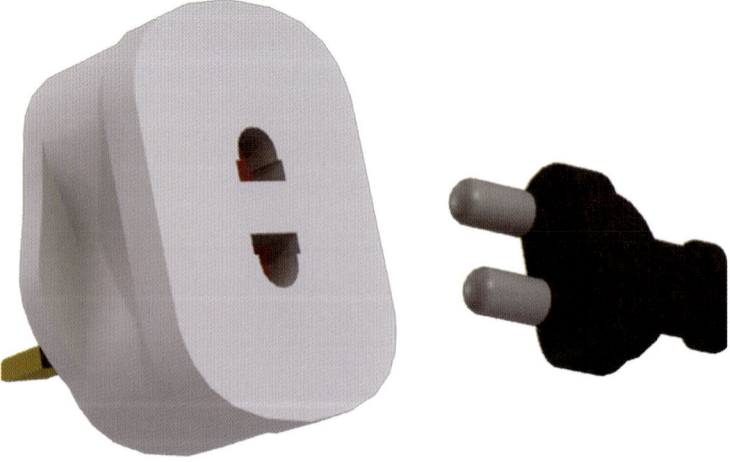

Plug and adapter

PANEL MOUNTING SYSTEMS

Mounting system options

When the power requirements are known the number and type of PV panels can be worked out. The best orientation and angle will also need to be established for the installation. There are three types of roof fixing options to consider, and other non-roof options are also possible. The roof options are to have an integrated in-roof array, a non-integrated on-roof array, or a flat roof array – assuming they have a flat roof available! Alternatively, ground mounted options or wall mounted options are also possibilities.

New house built with solar PV tiles

House with solar PV panels installed

Flat roof array

Ground mounted solar PV panels

Wall mounted solar PV panels

The pressure exerted on the roof needs to be calculated carefully. The load needs to include the weight of the fixings and panels and also the load must take account of the weather, in the form of extra weight such as snow and ice, and wind, as the panels will experience uplift during periods of high wind that can be considerable.

The load must take into account different weather conditions

ACTIVITY 23

If wind uplift is an issue, for a:

- pitched roof
- flat roof

discuss what you might fix the PV panel to (on the building) and to what depth, or what alternative methods might you be able to use?

Temperature issues

Temperature needs to be considered, as it affects the efficiency of the panel. Conversion of solar energy to electricity decreases as temperatures rise. How the array is fixed to the roof therefore becomes important in the design of the installation. Battens and framework must allow for ventilation at the sides, top and bottom of modules.

Integrated modules do not perform as well as an equivalent non-integrated array because the ventilation and temperature issue interferes with electricity production and reduces efficiency.

Converting solar energy to electricity decreases as temperatures rise

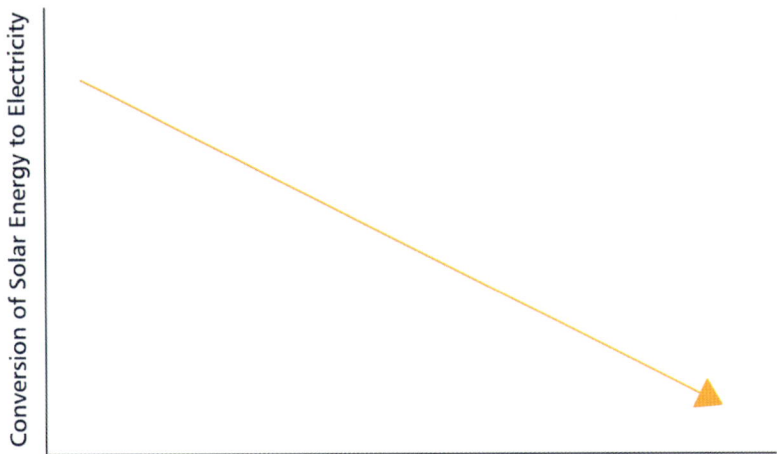

Graph showing relationship between converting solar energy and temperature

ACTIVITY 24

Above, it states that integrated panels are not as efficient as on-roof mounted panels. Please state why this is and how you would design in-roof panels (integrated) so that they were as efficient as on-roof mounted panels.

In-roof PV arrays

Domestic in-roof, or integrated PV arrays are usually PV slates and PV tiles. They can match existing tiles or slates and there is a wide choice of styles and colours available. They can be used for the whole roof area or be used over part of the roof.

PV laminates can also be used as an alternative, which are PV rigid modules without an aluminium frame. The system mountings consist of an aluminium frame screwed to the rafters and clamped to the PV modules. Flashing is provided for weatherproofing.

Aluminium frames, PV modules

Flat roof PV arrays

Flat roof PV arrays use substantial aluminium or galvanized frames. Careful attention is given to sealing and weatherproofing the roof. Troughs filled with ballast can also be used. Because the panels are frame mounted they can use optimum tilt and orientation and have good ventilation. The drawback is that the array must withstand considerable wind uplift. In areas where snowfalls are likely there are additional risks from localized snowdrifts building up under the modules and causing the weight load to increase. This can be mitigated by adding vertical screens behind the panels or raising modules so the snow blows away.

Flat roof PV arrays

Non-roof mounted options

If the weight or wind load would be too great for the roof it is possible to use flat roof techniques on the ground. This is often used for large scale PV power generation plants on unproductive land. In domestic situations it will require a large space in the garden. Ground mounted panels need the vegetation such as grass kept in check to prevent shading.

Non-roof mounted PV arrays

Vertically mounted

PV modules can be mounted vertically on façades of buildings, and made into a feature providing an attractive appearance as well as electricity. Vertically mounted panels do not receive as much sunlight as those orientated optimally, but an increase in the surface area will compensate for this loss.

Increase the surface area

ACTIVITY 25

Looking at the angle and orientation graphs (chapter 1, pp. 14–15), state what is the percentage performance of a vertically mounted façade collector when it is pointing:
1. Due south
2. Due south-east
3. Due west

Pole mounted

Pole mounted PV modules need robust foundations using wet concrete and screw anchors, as the uplift effect from wind can be very great, but maintenance is lower as there should be no problems with vegetation shading the panels. Pole mounted options are suitable for small arrays up to 600W and can be used with a solar **tracker**.

Tracker A mounting for a small array that pivots the array in one or two directions to follow both the Sun on its East–West path and solar altitude in the sky

Pole mounted PV modules

BATTERIES AND CONTROLLERS

Batteries

Batteries provide an alternative power source and lead-acid batteries provide the best compromise between cost, efficiency and longevity. Other battery options are being explored. Batteries do have a limited lifespan and require careful maintenance. Batteries can be connected together in series or parallel like the panels to provide sufficient power to meet requirements.

Batteries need to be placed where the temperature will not fluctuate too much, so are usually placed on racking. If the temperature variation is likely to fall outside a range of 0°C to 45°C then the housing should be insulated. However, never put any insulation on top of the batteries as potentially explosive gases need to escape!

Deep discharge batteries suitable for solar PV need to be used. Car and truck batteries are not suitable as they are designed for starting engines with a short burst of high current.

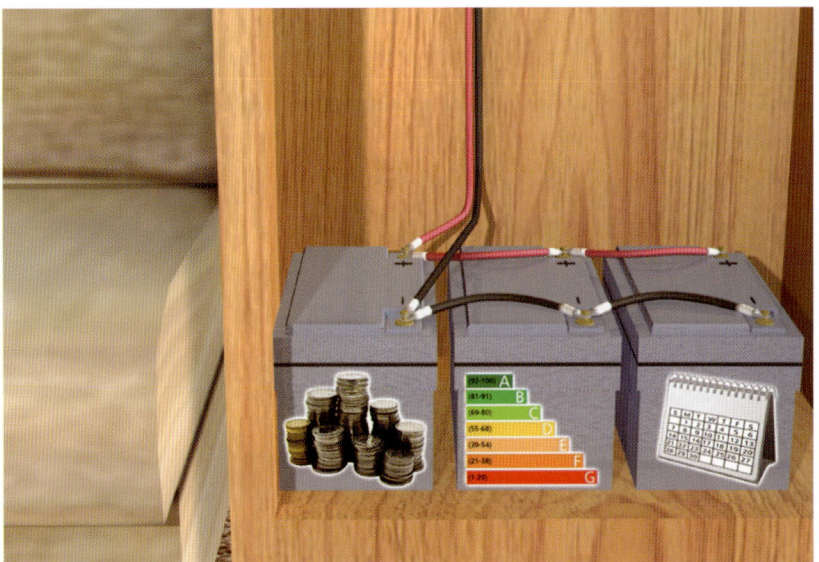

Batteries provide an alternative source

Cycling

Battery life span is measured in cycles – each time it is discharged and recharged it is a cycle. The expected life of a battery is usually given, for example 350 cycles or 1600 cycles. Deep cycle batteries are capable of deep discharge, sometimes being left discharged for long periods of time, such as in a holiday home. Battery life is improved if the battery is kept charged.

When calculating battery size and number the 'holdover' time needs to be considered. This is how long the battery supply will be needed to provide power before it starts to charge again. Usually this is taken to be between three and five days.

Battery life cycles - every time it is discharged and recharged it is a cycle

Controllers

A controller is needed for maximum battery life. The controller limits the amount of charge and discharge on and from the battery.

The load is the demand placed on the electrical system at any one time. This is used to calculate the number of PV modules required. The load controller electronically disconnects appliances and/or lighting when the batteries are fully discharged, i.e. flat, to prevent battery damage.

The **charge controller** limits battery charging when battery is fully charged – again to prevent damage.

Load and charge controllers can be combined in one unit and can display the current, voltage and power generated and stored in the battery. The charge controller is always placed between the PV array and the battery, as near the battery as possible.

> **Charge controller** A device to limit the charge received by a battery once it is fully charged, and is often combined with a load controller used to limit the electrical load put on a battery to prevent damage

Controller stops charging when the battery is fully charged

E-LEARNING

Use the e-learning programme to learn more about controllers.

ACTIVITY 26

Thinking about Battery Charge Controllers (BCC, which are also often called Load and Charge Controllers), what other features besides protecting the battery from under or over charge, might be useful or desirable? And how do most commercially available BCCs check that the battery is fully charged or discharged?

INCREASING THE SUNSHINE

Solar trackers

Solar trackers can move the panel or array to track the Sun across the sky, which can increase the volume of sunlight received by 15 per cent to 20 per cent in winter and up to 55 per cent in summer. They can control both the elevation and angle of the panel so the maximum solar radiance is collected during each day. The drawback is that solar trackers are expensive, and if they break down, the panel is likely to be positioned in a poor position until it is repaired.

Solar trackers are sometimes called **heliostats**, or can be called Heliostat Concentrator Photovoltaics and referred to as HCRV.

> **Heliostat** A mirror that tracks the movement of the Sun to reflect its radiation onto the PV cells/modules

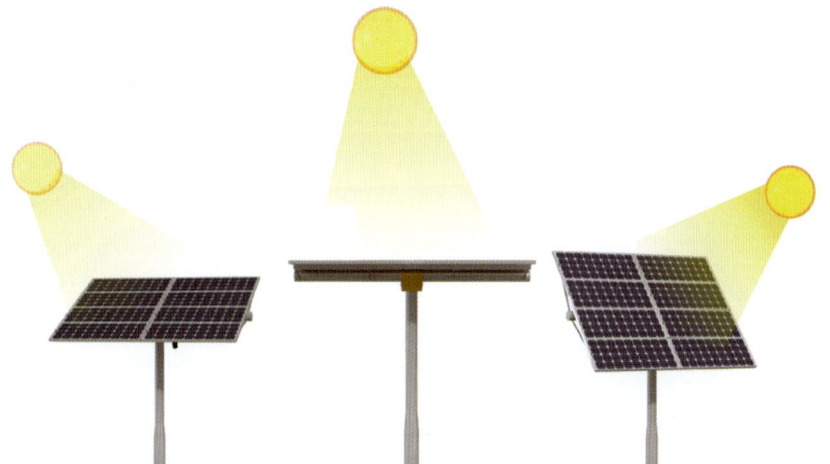

Solar trackers can increase the volume of sunlight received

ACTIVITY 27

Please suggest two methods for making solar panels track the Sun in the sky. Suggest an advantage for each type of system.

Concentrators

Concentrators increase the sunlight received by using mirrors or panels with **fresnel lenses** – which are used in lighthouses – to focus additional light onto the PV panels.

> **Fresnel lens** Use of this lens allows a way of concentrating solar radiation to focus additional light on a PV panel

In the UK and other northern European countries it can be cost effective to use fixed parabolic concentrating systems, and can reduce the number of panels required to provide the same level of generated electricity.

A drawback is that by concentrating the sunlight you get a corresponding rise in temperature, which also affects the efficiency of the panel. If using mirrors they must not dazzle other people at any time.

Fresnel PV panels tend to be larger than ordinary ones, and long term reliability has yet to be proved.

Concentrators can increase the sunlight received by using mirrors

STAND ALONE SYSTEMS

Layout with DC circuit only

In a simple stand alone circuit the main components will be a PV panel, controller, battery, DC distribution board and DC appliances. The whole will be connected by DC cabling throughout.

The advantage of using DC current only is more efficient use of the power provided, as there is no need to have a large inverter on standby all the time. However, there are losses associated with storing the energy in a battery which can be similar to inverter losses. Stand alone PV systems are cheaper to run in remote locations than using a petrol or diesel generator.

Cross section of a holiday mobile home, with solar PV panels

There should be a fuse to the battery or DC isolator depending on size and number of batteries, and the distribution board should have an isolator or main switch, possibly a double pole switch. If the PV panels are a distance from the controller and batteries – such as on the roof and the batteries in the garage – another isolating switch could be positioned near the PV panels.

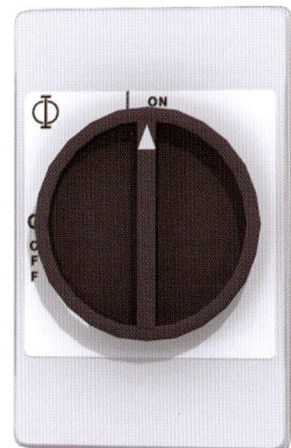

DC isolator

ACTIVITY 28

Where might you find information or guidance on sizing and specifiying solar DC (and for that matter solar AC) systems?

The future of electrical energy storage

Currently batteries are used for storage. In the future, we might use fuel cells to store the energy instead of batteries. Currently, they are too costly and ineffective for small scale systems.

Battery requirements are usually 12V for caravans and mobile homes, 24V for houses, and larger commercial systems will typically use a 48V battery store.

*Battery requirements are usually
12V for caravans and mobile homes*

Layout with DC and AC circuits

Inverters can be used with stand alone systems. Many stand alone systems operate at 12 or 24V DC current circuits, but the majority of household equipment is designed for 230V AC current in the UK. Using an inverter in the stand alone system allows householders to use ordinary equipment. DC cabling is needed up to the inverter, then normal AC cabling can be used.

Many stand alone systems operate 12 or 24V DC current circuits

Using DC and AC circuits

It is possible to have a stand alone PV system with both DC and AC circuits. Use of everyday electrical goods allows greater flexibility. Both circuits will need a separate distribution board. Dual systems may use the DC supply for lighting and AC supply for other appliances. It is important to have different plugs and sockets for each type of circuit to prevent connecting a DC appliance into an AC socket so as not to cause any accidents.

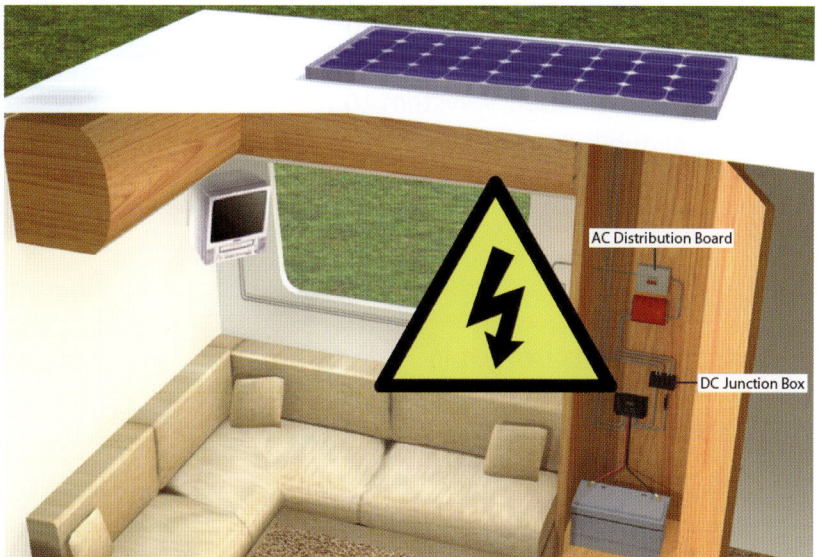

It is possible to have a stand alone system with DC and AC circuits

GRID CONNECTED SYSTEMS

Typical grid connected system

Grid connected systems do not usually have any batteries or battery charge and load controllers in their circuit, but pass the generated electricity straight to the inverter, via a DC isolator. There may be several DC isolators depending on the distance between the PV panels and inverter, for example one in the roof by the panels and one next to the inverter. The AC current then passes through an AC isolator to the generation meter and on to the distribution board. The distribution board is fed by both the generation meter and the import meter – both of which have to be approved by the local grid distributor. Distribution boards are also often called consumer units.

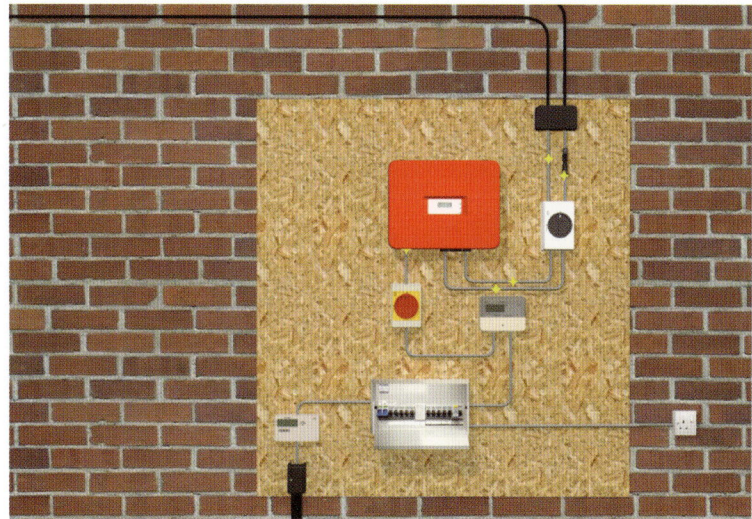

Grid connected systems don't usually have any batteries or battery charge

During daylight the PV modules generate electricity and supply the household needs. Any excess is taken by the grid. However, if demand becomes too great for the PV system to meet, then electricity is taken from the grid as well. At night all demand is met from the grid.

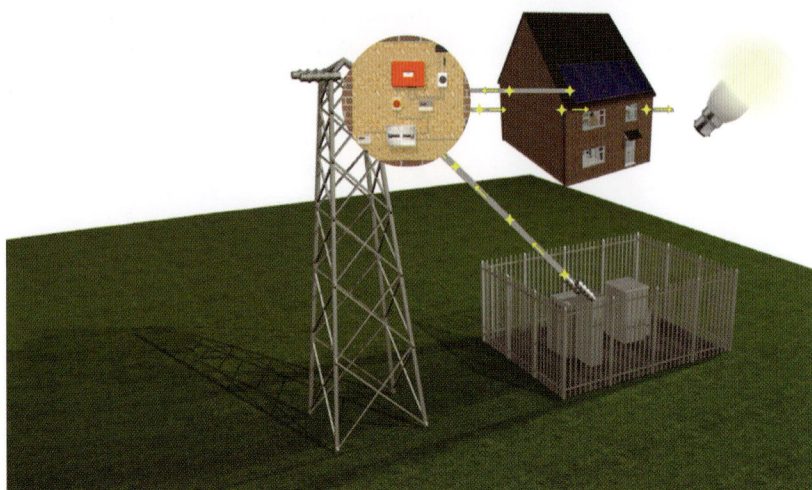

Any excess electricity generated is taken by the grid

Grid connected systems are constructed to turn off if there is a power failure in the local grid. The inverter cuts off the supply to protect anyone working on grid repairs inadvertently being electrocuted. Customers need to be aware that this situation may arise, and that having a PV system does not mean they cannot have their electricity turned off!

Grid connected systems are designed to turn off if there's a power failure

E-LEARNING

Use the e-learning programme to see examples of grid connected systems.

ACTIVITY 29

Your customer wants to be able to maintain the power supply to their house in the event of a power cut. What are the requirements you need to discuss with your customer and what budget and maintenance implications might this include?

STAND ALONE VS GRID CONNECTED

A comparison of a situation where there is a grid connected solar PV system which is being used to generate income through feed-in tariffs, and a household using a stand alone system which is also generating income from the **feed-in tariffs**, gives a clear distinction between the two types of installation.

Feed-in tariffs (FiT)
When an electricity supply company is obliged to pay a minimum rate per kWh for renewal energy

Item	Stand Alone System With Feed In Tariff	Grid Connected System With Feed In Tariff
Meters	Approved generation meter to obtain feed-in tariff	Various, depending on how feed-in tariff works
Batteries & Controllers	Yes	None
DC Appliances	Probably	None
Excess electricity generated	Stored in batteries	Fed into grid
Electricity supplied when dark or insufficient sunlight	Supplied from batteries	Supplied from grid
Isolators	DC and possibly an AC isolator	Additional DC and AC isolators
Size of PV Panels	Meets requirements	May be smaller than needs or far exceed requirements

Comparison between stand alone and grid connected systems

A stand alone system will have a charge and load controller and batteries, the grid connected system will not. Also the stand alone system will probably have at least some DC appliances and lighting, whereas the other will not.

Item	Stand Alone System With Feed In Tariff	Grid Connected System With Feed In Tariff
Meters	Approved generation meter to obtain feed-in tariff	Various, depending on how feed-in tariff works
Batteries & Controllers	Yes	None
DC Appliances	Probably	None
Excess electricity generated	Stored in batteries	Fed into grid
Electricity supplied when dark or insufficient sunlight	Supplied from batteries	Supplied from grid
Isolators	DC and possibly an AC isolator	Additional DC and AC isolators
Size of PV Panels	Meets requirements	May be smaller than needs or far exceed requirements

Comparison between stand alone and grid connected systems

The electricity generated in the grid connected system will go straight from the PV panels through an approved inverter to become a 230V AC system, but the stand alone system may use some or all as DC current, or feed some through an inverter to provide AC current for larger appliances.

Item	Stand Alone System With Feed In Tariff	Grid Connected System With Feed In Tariff
Meters	Approved generation meter to obtain feed-in tariff	Various, depending on how feed-in tariff works
Batteries & Controllers	Yes	None
DC Appliances	Probably	None
Excess electricity generated	Stored in batteries	Fed into grid
Electricity supplied when dark or insufficient sunlight	Supplied from batteries	Supplied from grid
Isolators	DC and possibly an AC isolator	Additional DC and AC isolators
Size of PV Panels	Meets requirements	May be smaller than needs or far exceed requirements

Comparison between stand alone and grid connected systems

The grid connected system will have additional AC and DC isolators.

Item	Stand Alone System With Feed In Tariff	Grid Connected System With Feed In Tariff
Meters	Approved generation meter to obtain feed-in tariff	Various, depending on how feed-in tariff works
Batteries & Controllers	Yes	None
DC Appliances	Probably	None
Excess electricity generated	Stored in batteries	Fed into grid
Electricity supplied when dark or insufficient sunlight	Supplied from batteries	Supplied from grid
Isolators	DC and possibly an AC isolator	Additional DC and AC isolators
Size of PV Panels	Meets requirements	May be smaller than needs or far exceed requirements

Comparison between stand alone and grid connected systems

The size of a solar PV system will normally be related to the expected maximum load in a stand alone system. The number of PV panels in grid connected systems will not be related to expected load but to how much the client wants to generate to assist, meet or exceed their requirements and use feed-in tariffs.

Item	Stand Alone System With Feed In Tariff	Grid Connected System With Feed In Tariff
Meters	Approved generation meter to obtain feed-in tariff	Various, depending on how feed-in tariff works
Batteries & Controllers	Yes	None
DC Appliances	Probably	None
Excess electricity generated	Stored in batteries	Fed into grid
Electricity supplied when dark or insufficient sunlight	Supplied from batteries	Supplied from grid
Isolators	DC and possibly an AC isolator	Additional DC and AC isolators
Size of PV Panels	Meets requirements	May be smaller than needs or far exceed requirements

Comparison between stand alone and grid connected systems

CHECK YOUR KNOWLEDGE

1. **Identify the labelled areas of the stand alone PV installation shown.**

1.

2.

3.

4.

2. **Match the correct type of mounting system with the images shown.**

Mounting System	Image
Solar PV tiles	

Solar PV panels

Flat roof array

Ground mounted
solar PV panels

Wall mounted
solar PV panels

3. **Identify the labelled areas of the grid connected PV installation shown.**

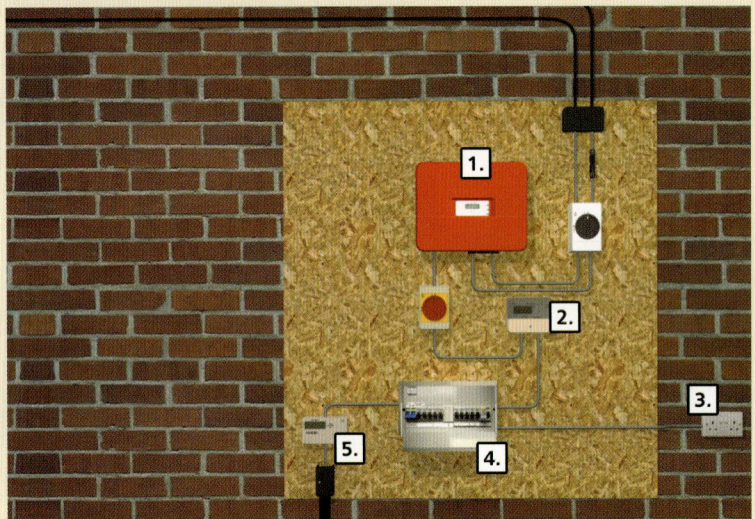

1.

2.

3.

4.

5.

4. **Look at the highlighted cables shown. What type of cable would you expect to find there?**

_____

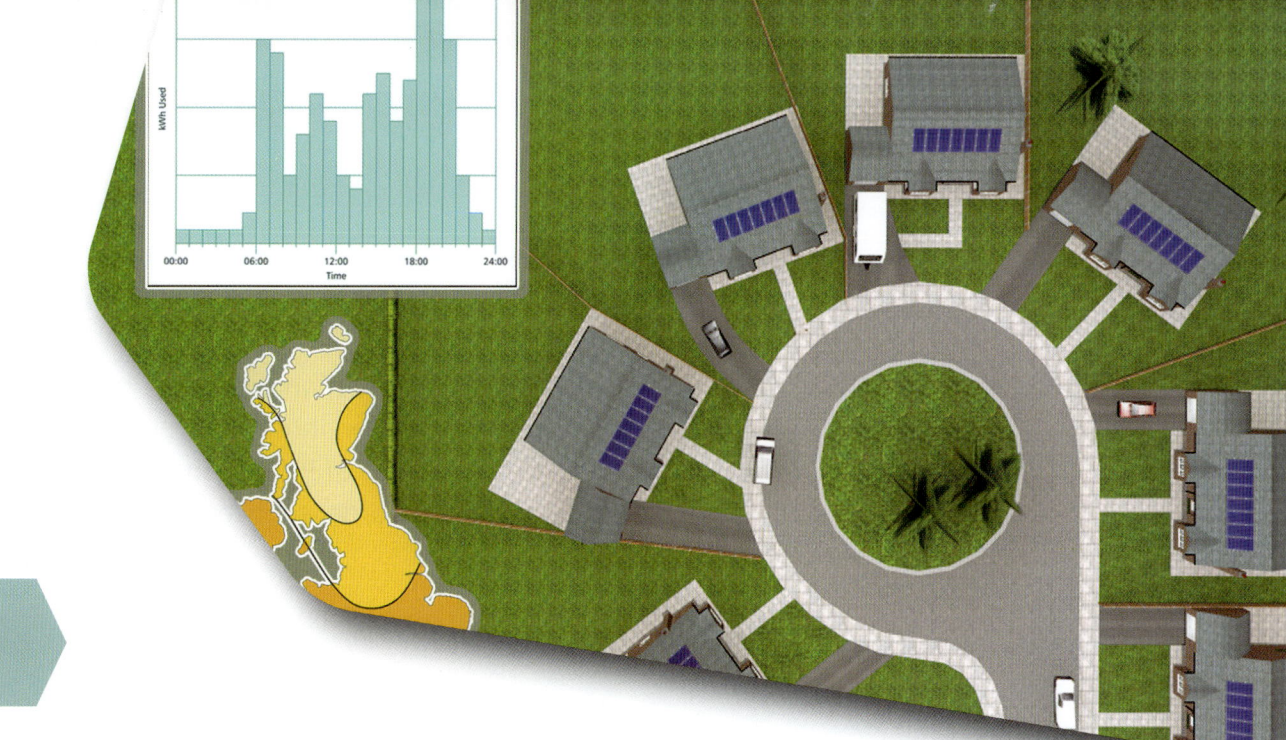

Chapter 6

SYSTEM DESIGN

LEARNING OBJECTIVES

By the end of this chapter you will be able to:

- Describe the fundamental design principles used to determine the best position for a solar PV system

- Describe the fundamental design principles used to determine the size of the module array to meet the household requirements

- Calculate the requirements of a household

- Calculate the nominal power (KWp/m^2) of a given product

- State the importance of shadow on the installation

- Describe the advantages and disadvantages of a solar tracker within a solar PV system

DESIGN PRINCIPLES

This chapter looks at the design process and how to gather the information needed in order to make an informed choice on the best solar PV installation for the location and budget available.

There are four areas where information is needed: how much energy from the Sun is received at the location, how much energy the household uses – or proposes to use, what size of PV system will supply that need, and what options do they have in order to meet the needs and available budget.

Satellite view of houses with solar PV panels installed

Measuring the house's electric consumption

Solar PV panels installed on a rooftop

Ground mounted solar PV panels changing angles following Sun's position

To design an optimum PV system you need to:

- Know how much solar energy is available for that location
- Know the household electrical energy requirement
- Calculate the size of the PV modules required to meet requirements
- Determine the options available within budget

There are different factors to consider when installing solar PV panels

SITE ASSESSMENT

What do you need to know?

In order to design a good PV system which will produce the maximum amount of power over the years at a specific location, you first need to know how much energy from the Sun is available. This includes knowing the direction and the annual variation for the particular site of the solar energy. Potential shadows need to be identified as well, as these can have a large effect on the efficiency of the system.

Roof top with solar panels installed

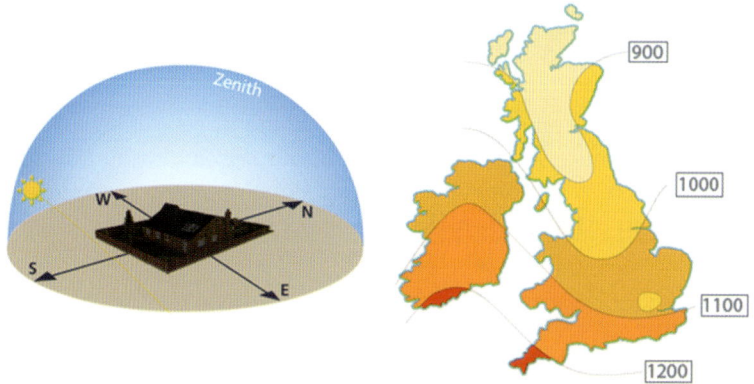

The Sun's direction is an important factor to consider

Angle of tilt

As a general rule PV panels in the UK should face south at an angle of about 35° for maximum efficiency. Other options are possible which work at a slight loss in efficiency. The loss can be compensated for by having a proportionally larger area of PV panels than would be required if everything was perfect. The diagrams show typical loss of efficiency according to roof pitch.

Pitch depends on latitude; 35° is an average for the UK. It would be slightly higher in Scotland and lower in Cornwall. Pitch helps the rain to keep the PV panels clean, and in areas with snowfall a pitch of at least 45° is needed to prevent snow build up. A pitch above 21° is recommended for self cleaning from rain, keeping dust build up down.

> **Roof pitch** Angle of the roof slope, which will affect the inclination of any on-roof or in-roof PV system

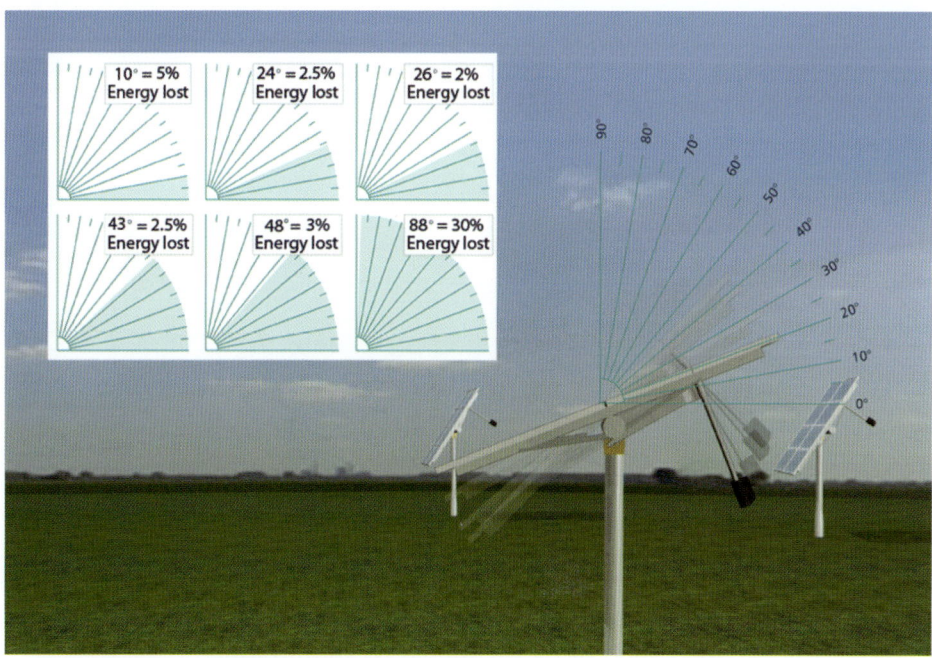

For maximum efficiency, a panel should be placed at a 35 degree angle

ACTIVITY 30

This section talks about angle of tilt for self-cleaning of solar panels. Suppliers and commentators make various statements about this angle and also suggest other methods such as self-cleaning coatings. What should you consider with regards to cleaning matters when thinking about solar PV panels?

Orientation

Orientation ideally should be towards the south, but like angle of tilt, other directions are possible without great loss of efficiency. Facing east or west can provide up to 80 per cent of optimal performance, although installations on north-facing roofs are not recommended. Much of the solar energy the UK receives is in the form of diffused light, so although direct sunlight is best, some light is received during daylight no matter which direction is faced. Orientation is also referred to as the azimuth angle.

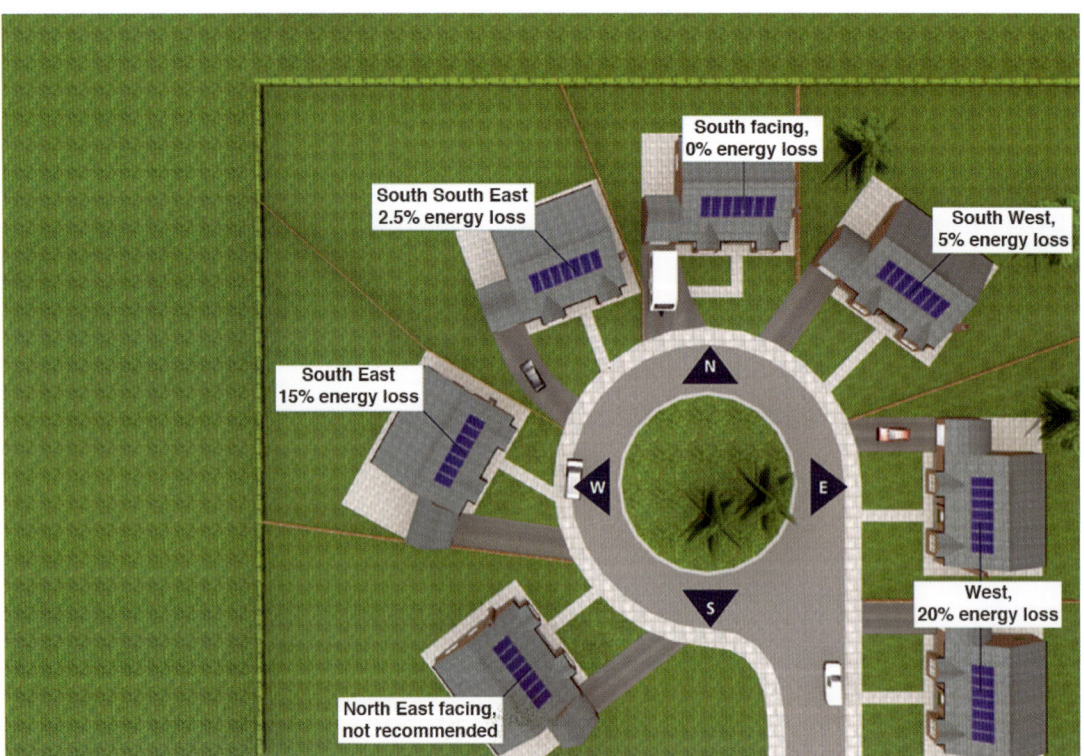

For maximum efficiency, ideal orientation of panels should be south-facing

ACTIVITY 31

If the PV array is to be located east- rather than south-facing and the system only collects 80 per cent of the energy of a south-facing system, how much does the array have to be increased in size to compensate for the 20 per cent loss of energy? Therefore, how much surface area would be required of an east-facing system to match a 20m^2 PV array facing directly south?

Solar radiation

The solar energy received at any one place on Earth is determined by its latitude and distance from the equator. It is also the result of the climate – or weather. If the house is by the coast the skies are likely to be clear reasonably often and therefore receive more sunshine than say, in a mountain valley where there is high rainfall from dark clouds. The typical amount of annual direct sunshine, light cloud cover and dark cloud cover can be found from weather records. The annual solar radiation is measured in kWh/m^2/annum. Seasonal variations will be included in this figure.

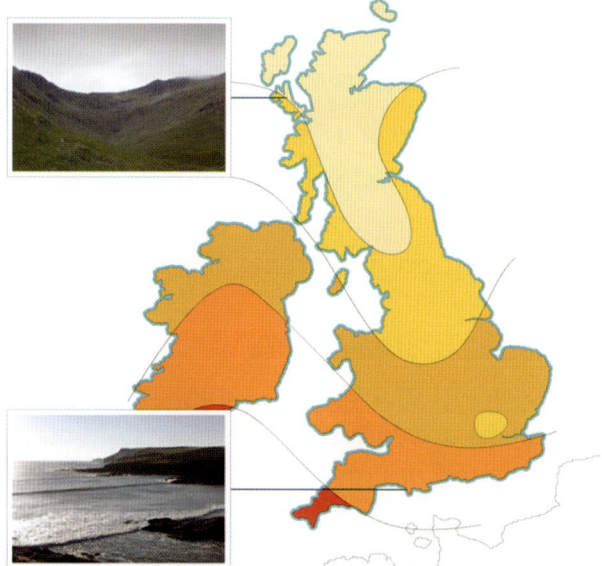

Coastal areas are likely to receive more sunshine than a mountain valley for example.

Each day gives a variable amount of solar radiation: lowest at dawn and dusk, highest at midday. The seasonal variation also affects solar radiation; in summer the Sun is at its highest but in winter, even at midday, it is low in the sky.

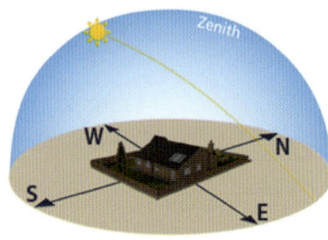

Summer time, 21 June, 4 am

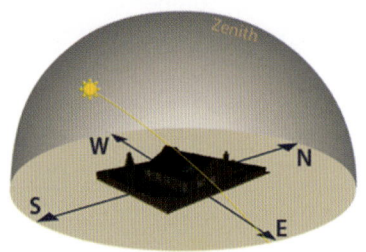

21 September, 21 March 6.30 am

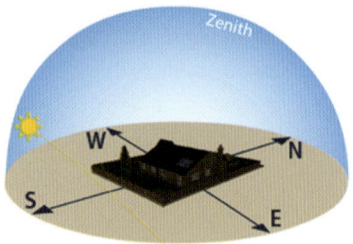

21 December 8.30 am

ACTIVITY 32

Is solar radiation at its highest on the equator? If not, why might this be? What percentage of the Sahara Desert is needed to be set aside to provide power for Europe? And why is this comment politically sensitive?

Shadow

Shading is critical for PV panels because the cell within the panel receiving the lowest amount of solar radiation determines the output for the whole panel. As an illustration think of a garden hose 50m long; someone treading on it prevents the flow of water along the whole hose.

The proposed location of panels for a PV installation needs to be examined for shadows from nearby chimneys, buildings, trees, TV aerials and vent pipes. All are common causes of shading and should be accounted for before the design of any installation. A system can tolerate a little shading early or late in the day but should ideally never be shaded between 10 am and 4 pm.

Shadows are critical - the lowest amount of radiation determines the panel's output

ACTIVITY 33

Why is solar PV so much more sensitive to shading issues than solar thermal installations? Does this mean that we should always fit solar thermal in preference to solar PV systems?

Shadow and Sun path diagrams

The path the Sun takes in the sky can be plotted on a graph. The graph shows the height of the Sun in the sky and the angular difference from due south. During the summer the Sun is at its highest and in the sky for longest – indicated by the length of the line. The space underneath the curved line shows the solar radiation available – assuming the day is cloudless!

In winter the days are much shorter and the Sun does not reach very high above the horizon. For stand alone systems the winter performance is a high priority as this is when demand will be highest.

Between these two lines all other dates can be accommodated – for example, halfway between the maximum and minimum will be the Sun's pathway on 21 March and 21 September.

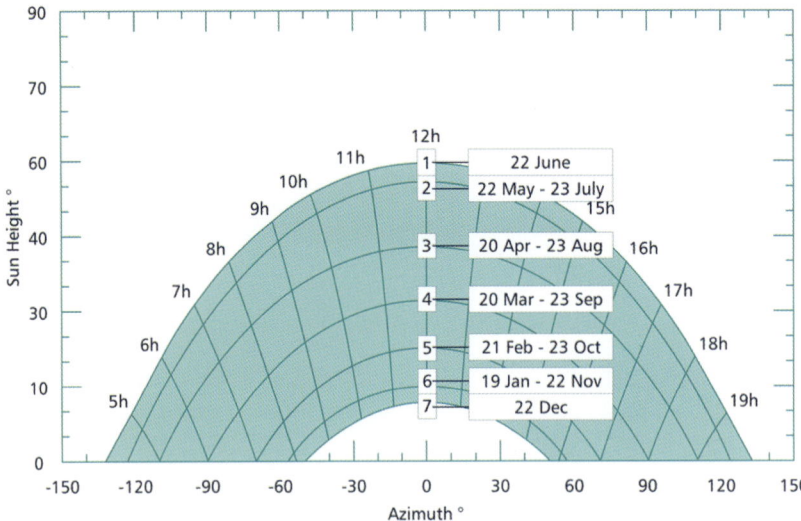

A graph showing the Sun's path across the sky

Sun path diagrams
These are used to calculate available solar radiation and identify any shadow that may affect PV efficiency, or identify alternative sites, during the design stage

---

FUNCTIONAL SKILLS

Sun path diagrams are available for any latitude and can be downloaded from the Internet.

The effect of shadow can be calculated from a Sun path diagram. The angle of the object causing the shadow can be measured and plotted onto the graph and the available solar radiation calculated. For example, a chimney nearby may provide shade where the PV array is to be positioned in the late afternoon between March and September, but not affect the array during winter between September and March.

However, a large factory next to the house on the southern side may severely restrict direct sunlight most of the year, only allowing direct sunlight during the middle of the day in midsummer, when the Sun is high enough to clear the building. PV panels may need locating elsewhere in this situation – such as in the garden away from the factory shadow.

In the UK there is a lot of diffuse solar radiation available so although PV panels would not function at their optimum performance, they would still produce electricity during shaded periods.

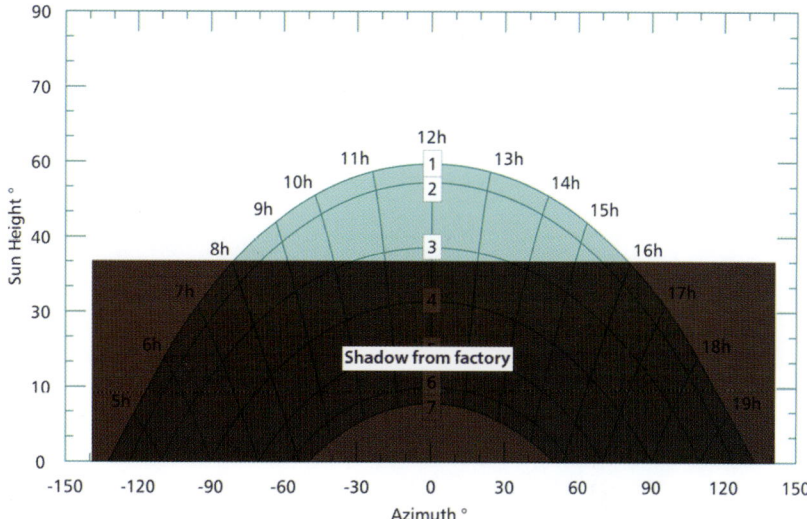

Graph showing how a factory's shadow can effect how much radiation a house receives

ACTIVITY 34

In this section it says that the UK has high levels of diffuse solar radiation. Do you think the UK would have more or less diffuse than a similar latitude in eastern Germany?

Shadows, hot spots and bypass diodes

Depending on how the array is connected, shading on a single module will affect the performance of the whole array. It is also possible for a cell within a panel to be shaded by something small, such as a leaf stuck by rainwater. A 'hot spot' may then develop which can destroy the cell, and therefore destroy the performance of the whole panel. To solve this potential problem panels can be provided with 'bypass diodes', which provide an alternative current pathway and divert current away from a shaded cell.

Panels made from amorphous silicon composed of strips are less affected because shading objects don't usually cover the whole length of a strip and therefore only affect the covered proportion of that strip. It also means amorphous strips are less likely to develop 'hot spots'.

A leaf stuck by rain water can also cause a shadow

Bypass diode

ACTIVITY 35

Some solar panels are being developed that have a bypass diode built into every cell. What effect might this have if the panel is partially shaded?

E-LEARNING

Use the e-learning programme for more information on the site assessment process.

CALCULATING HOUSEHOLD REQUIREMENTS

Calculating requirements

Each household has different demands for electricity. Installing a PV system requires an estimation, and preferably an actual measurement, of how much they use, and how they use it. Without this information the size of the PV array is difficult to calculate.

The purpose for installing a PV system needs to be established: were they going to install both solar PV and solar thermal systems, and was there a possible lifestyle change involved as well? All have a bearing on this process called a **load analysis**.

> **Load analysis** A method of calculating household electricity requirements when designing a PV system

The household electricity demands must be calculated

Reducing the existing load

The existing efficiency of the household needs to be investigated before designing a PV system:

- Is the house well insulated?
- Are the appliances energy efficient?
- Do they leave appliances on standby?
- What is the purpose of the PV system?

These questions need to be asked whatever type of system they have requested. Stand alone systems need to be as efficient as possible to prevent running out of power. Grid connected systems need to be as efficient as possible to maximize returns from the feed-in tariff. By eliminating unnecessary electrical consumption before a PV system is installed is good practice, and may require habits and lifestyle changes.

The existing efficiency of the household needs to considered

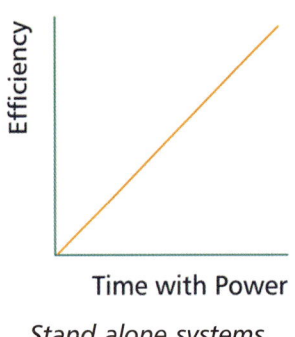

Stand alone systems

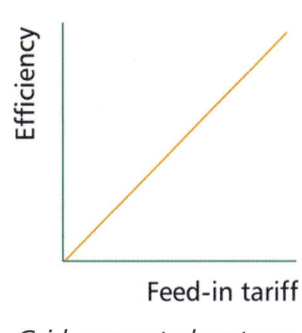

Grid connected systems

ACTIVITY 36

In this section it says that the insulation levels might affect the electrical energy consumption. In light of insulation affecting heating provision, discuss this and also, will solar PV have much effect on the heating provision in any electrically heated property? Therefore, how might fitting solar PV affect heating consumption?

Measuring the load

Power consumption within a household can be estimated, using the given wattage of individual items or tables, but it is much better if actually measured. Plug-in kWh meters can be used for each plug-in appliance to be monitored. Lights, heating and cooker usage is more difficult to measure but can be done.

Whole house energy meters are available, for example an 'Efergy' wireless electricity meter that can be clipped onto cables (meter tails) between the electricity meter and the consumer unit. **Smart meters** provided by electricity companies also give whole house consumption. Meter readings from utility bills can also be used. Measuring the load should be carried out over a period of time, such as a week or month rather than just one day.

Smart meters can measure the house's consumption as a whole

Once the load has been measured a load management graph can be plotted. Heavy usage will be clearly shown as peaks. These may be because the washing machine, vacuum cleaner and oven are often used simultaneously. For stand alone systems, spreading the load more evenly during the day will reduce peak demand, and therefore the size of required battery store to meet that demand. For grid connected systems, it is good practice to encourage better management of electrical loads by the occupants so as to minimize bills and climate change.

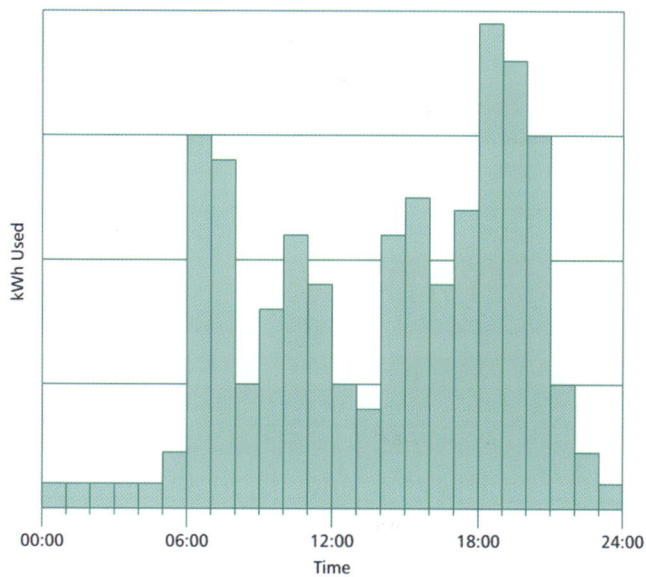

Load management graph

Calculating future load

Once there are accurate figures giving household demand for electrical power, the daily, weekly and seasonal variations in electrical demand can be studied and decisions made. Peaks should be eliminated if possible, by spreading the time when items are used. If this is not possible then knowing a peak will occur is being prepared, much like power companies anticipate half time during major football matches so there will be sufficient power available when millions of homes turn on their kettles.

Some items have a start up peak such as fluorescent tubes and fridge compressors. The item could be changed or the household behaviour changed to reduce their frequency, such as reducing the times the fridge door is opened.

Computer programmes are available to help work out future usage based on the figures gathered. Additional items will be added to the household load with the PV system, such as the inverter.

At the end of the process there will be an annual household requirement, measured in kWh, and the peak demand that needs to be met.

Once electrical household demands are calculated decisions can be made

DETERMINING SYSTEM SIZE

Meeting requirements

Once details of daily, weekly and seasonal variations in solar radiation are available for the location, and the household requirements in probable energy demands over the year have been measured, the size of the required system can be calculated. The reason the household wants solar PV also needs to be taken into account.

The size of an array will depend on whether the system is grid connected or stand alone. Sizing stand alone systems is more critical than for grid connected systems because the number of days running off batteries needs to be factored into the calculation.

For grid connected systems, within budget and available space, the bigger the system the better, as more energy means more returns from the feed-in tariff and more renewable energy onto the national grid.

20m² of PV panels generates 50% of household electricity

ACTIVITY 37

Assuming money is no issue, what would normally be the main factor affecting array size:

1. in a grid connected PV system?
2. in a grid disconnected PV system?

Panels

FUNCTIONAL SKILLS

As a rough estimate a system of about 20m² of PV panel is sufficient to deliver around 50 per cent of a typical household's electricity requirement in the UK.

Maximum power point (MPP) Point on the current-voltage curve where maximum power is produced

PV module specification sheets give many details about the panel, including optimum performance of voltage and current outputs at the **maximum power point**, shortened to MPP. These figures are rarely met in practice. The proposed orientation and tilt angle of the PV array will affect power output, as will the panel type.

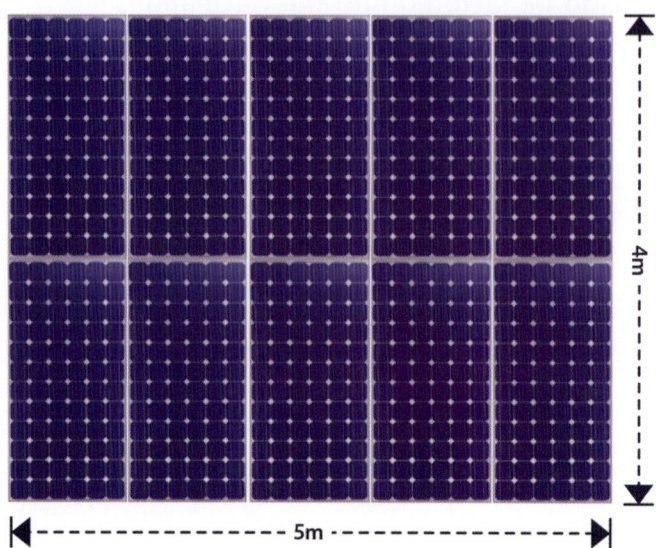

PV module specification sheets can give details such as optimum performance

ACTIVITY 38

A typical Brtitish household uses about 4500kWh of electricity per annum. If the household fits a 20m² system and the solar array is ideally mounted with an annual insolation rate of 1000kWh/m² on a south-facing roof at a 35° inclination and no shading, assuming the solar PV system supplies half of the household's electricity requirements, what is the efficiency of the panels and what type of PV panel has probably been fitted? What surface area would be required to make the same energy if a 7.5 per cent efficient thin-film panel was used instead? Finally, will the typical electricity consumption change significantly from household to household across Europe?

Calculating size/estimated costs can be done using computer programmes

The calculation

Computer programmes are available that calculate size and estimated cost of PV systems to provide specified energy requirements in a given location and climatic conditions.

Solar tracking

Solar tracking can increase the efficiency of an array by up to 15 per cent to 20 per cent in winter and up to 55 per cent in summer. Tracking can be in one or two dimensions, following the orientation and/or the angle of the Sun above the horizon during the day.

Panels need to be free standing to have solar tracking, not fixed on or in a roof. Solar trackers require maintenance, unlike PV panels which are effectively maintenance free. If a tracker breaks down in the wrong position, for instance at the end of the day, then on subsequent days PV performance will be greatly reduced.

Efficiency of solar panels can be increased by solar tracking

ACTIVITY 39

In this section it says that PV panels are effectively maintenance free. Please discuss this comment and explain why PV panels are said to be effectively maintenance free.

DETERMINING OPTIONS

In roof or on roof?

If the customer is having a new roof, installing integrated tiles or slates would make sense. PV integrated roof tiles or roofing slates and shingles are available in different colours that can form part of the weather-proof membrane or structure. There must be sufficient ventilation available in the roof space to prevent overheating.

If they are not replacing a roof, the most cost effective option is to use standard PV modules in a frame bolted onto the existing pitched roof or a tilted frame on a flat roof. Modules can be arranged in many ways, especially to avoid shading. PV panels can be mounted vertically on walls in frames, but will lose some performance.

A house with PV integrated roof tiles or roofing slates

Using PV modules is more cost effective than replacing the roof

Comparative costs

The PV material selected will determine the overall cost of the system. PV materials that are more efficient tend to be more expensive, but require a smaller area.

Hybrid PV materials are more efficient, either as in roof modules or on roof panels, giving an output of 1kW per 6–7m^2. Thin film amorphous roof tiles, however, require 15m^2 to 18m^2 to produce the same amount.

Hybrid PV

Thin film amorphous

Additional costs

Additional costs when installing a PV system may or may not be included in the break down of the quote; however, these extras should be explained to the customer.

These include scaffolding, making good internal and external work, connection agreements with the local distribution network operator (DNO), lightning protection, AC connection and a display meter.

DNO Distribution Network Operator. The local area grid operator

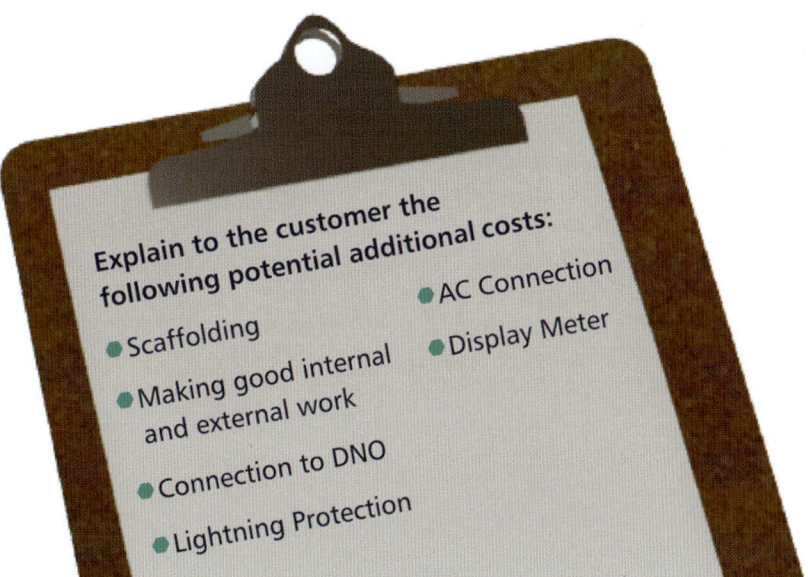

Additional costs should be explained to the customer

Grid connected system costs

Grid connected systems can be installed to meet some or all of the customer's power needs, or to generate electricity in order to receive an income through the generation tariff and feed-in to the local area grid.

DNO approved generation meters usually have electronic MPP trackers included, so the system will make as much energy as possible.

UK AND INTERNATIONAL STANDARDS

The inverter needs to be the right size for the system and DNO, and in the UK, the system must conform to G83/1, the relevant section of the building regulations (or the relevant local connection protocol of the region).

The price of selling back excess electricity may not be the same as the cost of electricity used from the grid. There may be a need to re-educate users how and when to use electricity to keep their own costs down and become more efficient. Output can be improved by adding a solar tracker where possible. Replacing inefficient electrical appliances would also maximize return on their investment.

Building regulation guides

ACTIVITY 40

Payments from feed-in-tariffs (FiT) tend to be much greater than the purchase price of electricity. For example, it is not unusual for the FiT payment (cents or pence/kWh) to be two to three times greater than the purchase price of electricity from the local utility. As a businessman or woman selling a solar PV system to your customers, what reassurance could you offer the client to demonstrate that the utility is fully committed to paying the FiT payment for the agreed lifetime of the incentive?

Stand alone system costs

Stand alone systems need to have the correct size of array – otherwise they risk cutting power to the occupants at times of peak demand. However, if the occupants are prepared to forego power for some of the time, then a stand alone system array can be smaller and therefore cheaper.

Hybrid system Use of more than one renewable source of energy for a household, e.g. solar PV and small wind turbine

Some households alleviate this by adding a wind turbine to create a **hybrid system**, which could work well providing they live in a suitable area. In Britain, solar works better in summer and wind in winter, so if are both available, they can be ideally matched.

Stand alone systems must have the correct size array

ACTIVITY 41

Should stand alone PV systems also receive FiT payments similar to grid connected PV systems?

CHECK YOUR KNOWLEDGE

1. **What is the estimated loss of efficiency for the different roof orientations shown in the diagram? Select from the options listed here.**

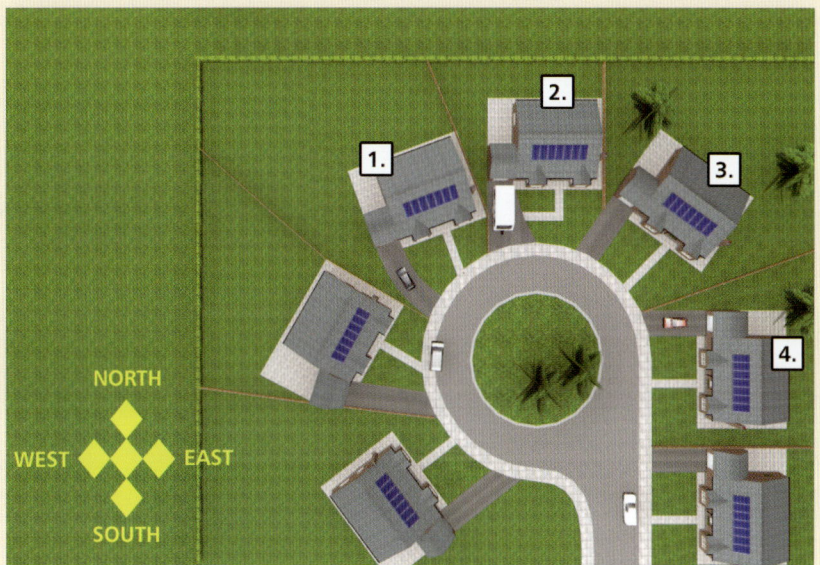

☐ a. no energy loss

☐ b. 2.5% energy loss

☐ c. 5% energy loss

☐ d. 8% energy loss

☐ e. 15% energy loss

☐ f. 20% energy loss

Chapter 7

PRE-INSTALLATION

LEARNING OBJECTIVES

By the end of this chapter you will be able to:

- Describe the pre-installation checks that must be carried out

- Identify the authorization required for the work to proceed

- Identify the requirements for inspection and testing of existing electrical installations

- Describe the arrangements possible for the arrays of solar panels (single and multiple strings)

- Describe the importance of roof strength for on/in roof installations

- Identify the internal siting of key components

WHAT DO YOU NEED TO KNOW?

Installation checks
Installation checks are made to ensure the installed system first complies with the original design specification for the customer, and then checked for any obvious defects

The purpose of pre-installation checks is to make sure installation work goes according to plan and the intended design can be achieved. It also gives the last minute breathing space to do a final check that the customer is happy with what they will be using in future.

Carrying out the necessary measurements before commencing installation

UK AND INTERNATIONAL STANDARDS

Authorization must be confirmed, the existing electrical installation requires checking in accordance with Part P of building regulations, and a final check that the building itself is OK for the extra weight and pressures.

Additional services required for the installation need to be arranged, such as scaffolding if required, and how the installation will proceed. Finally, check that the delivered materials for the PV are all present and fit for purpose.

Pre-installation checks are to ensure work goes according to plan

Materials needed to commence installation

WORK AUTHORIZATION

UK AND INTERNATIONAL STANDARDS

Pre-installation is the time to check that the installation will meet compliance requirements with building regulations and planning permission where needed.

Most domestic installations do not require planning permission, but they do need to be clear they are exempt. Permissions may change so it is best to check with the current planning permissions, which will usually be available on the Internet.

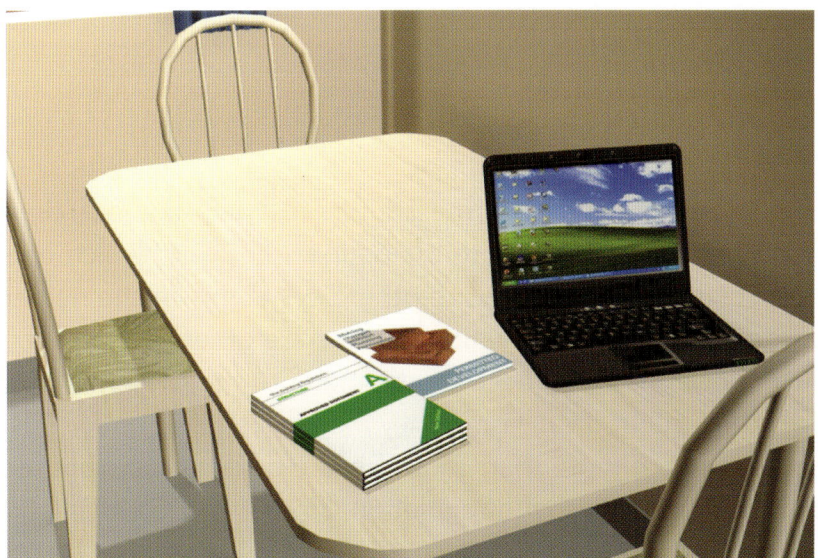

The majority of domestic installations do not need planning permission

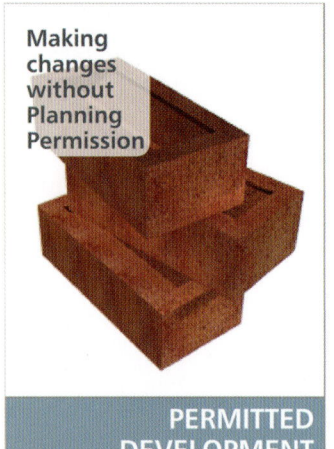

Making changes without Planning Permission

PERMITTED DEVELOPMENT

If in doubt use the General Permitted Development Order (GPDO)

If planning permission is needed it is frequently based on a need to blend in, and can insist on an installation having integrated tiles or slates or a free standing array rather than modules on a roof. Better to find out now rather than when the system is installed!

It is worth noting that most of the UK applies some form of the General Permitted Development Order or GPDO. This indicates that if the building is not listed or in a conservation area and the solar panels are less than 200mm above the roofline, then the installation is 'permitted development' which means no further permission is required. However, if in doubt, always check.

ACTIVITY 42

We have again looked at specific planning permission issues that apply only in the English and Welsh jurisdictions in the above section. This is because it is better to provide one example and permissions, regulations and standards regularly change both within and between regions. Can you think of ways to design PV systems so that they are likely to comply across many regions? List the features to ensure PV systems will comply with most local permissions, regulations and standards.

Supply to the grid

Amp (Ampere, symbol A) The unit to measure the flow of electrons forming the electric current

When a PV system feeds into the grid, it links with the local area grid and does not plug straight into the national network, which operates at a very high voltage. Therefore other PV systems in the neighbourhood could affect the installation. If there are other PV systems in the locality, prior application needs to be made to the DNO and they calculate if the local grid can cope. They may make a charge for this. This is also necessary if more than 16 **amps** are going to be generated by the system. It complies with regulation G59/1.

Houses with PV systems feeding into the grid

If there are no other PV installations and less than 16 amps are going to be generated, then the DNO must be notified before commissioning and all documentation must be submitted.

The inverter must always be approved by the DNO for all grid connected systems.

DNO approved inverter

ACTIVITY 43

National grids have up until recently been designed with the power generated at large centralized power stations and the power distributed to many millions of local 230 or more volts demand points such as houses, factories, schools and hospitals. Why does PV mounted on houses and other buildings change all of this?

STRINGS AND ARRAYS

The number of PV modules in a system depends on the required power. The modules may be connected in series or parallel. There may be just one string forming the array in a simple circuit, but more often there are several strings joined together to make the array. Each string requires a DC isolator.

PV modules may be connected in series or parallel

Simple one string array

This simple grid connected system is made from one string forming the array, with one inverter linked to the generation meter. DC and AC isolators are positioned either side of the inverter.

This system is made of one string forming the array; Single string array

Two string array, one inverter

There are two strings in this arrangement, each with a DC isolator, feeding into the same inverter. The inverter is then connected to the generation meter via the AC isolator. Several strings can feed into one inverter.

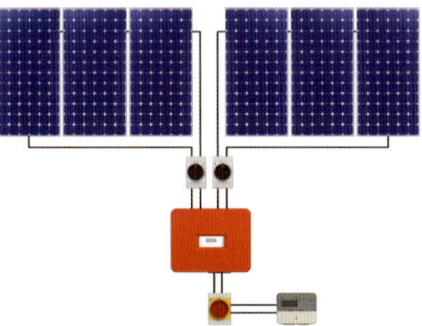

Two string array, one inverter

Two string array with two inverters

The two strings in this arrangement each have their own DC isolator, but also connect to their own inverter. This situation indicates the strings are positioned some distance from each other and would require an excessive length of DC cable to link them to the same inverter. The loss of energy in the DC cable would exceed the power used by running two inverters. Each inverter is then connected to an AC isolator and both feeds are connected to the single generation meter. Generation meters can receive current from several inverters.

Two string array with two inverters

Three string array with two inverters

Many combinations are possible with the modules and strings, and sufficient isolators need to be included. All the equipment, including inverters, cables and generation meter have to be of the correct size for the expected maximum power.

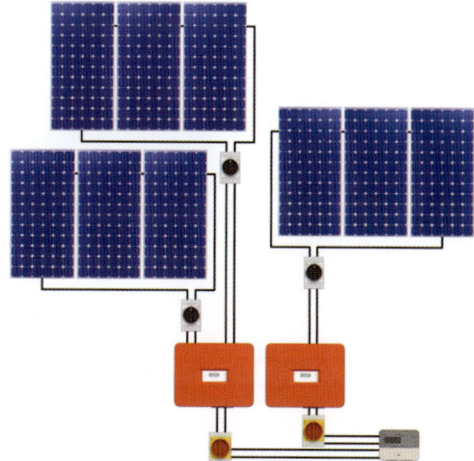

Three string array with two inverters

Export meter Meter that records all the electricity generated by the PV array and distributed to the grid

E-LEARNING

Use the e-learning programme to learn more about strings and arrays.

ACTIVITY 44

In different countries, various layouts are used for generation or **export meters** feed-in tariff payment systems. In the diagram shown, an export meter FiT layout is shown as would be used in Germany where the tariff is paid out in exported energy. In the UK, a generation meter is used, as the tariff is paid out in generated energy. Where would the generation meter need to be on this diagram? And would the export meter still be needed?

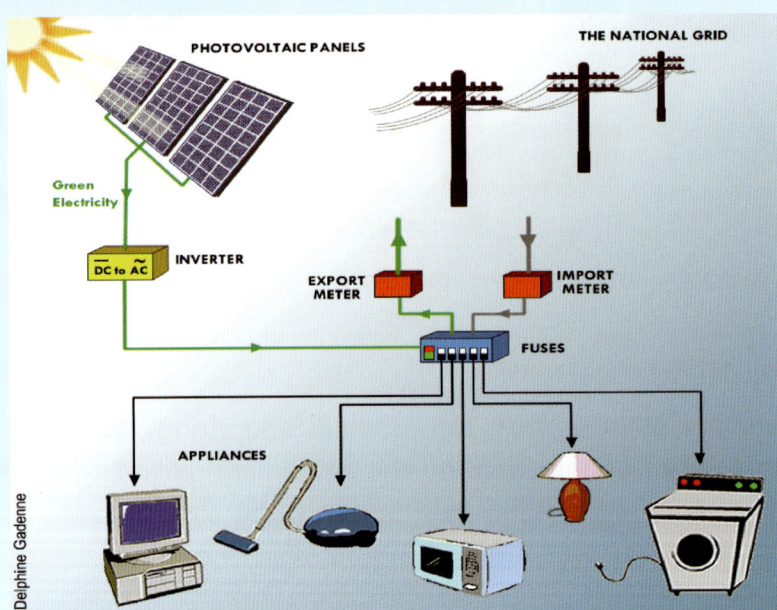

Delphine Gadenne

CHECKING THE LOCATION

Pre-installation checks can include another check of where the modules will be located. The location can be checked for shadows, and checking the optimum position for maximum collection capacity.

Incorrect location

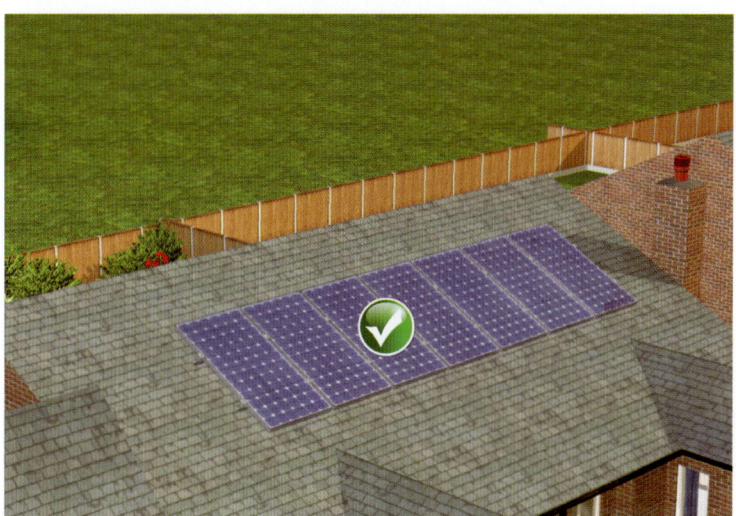

Correct location

Checking roof locations

A roof and the whole building needs to be checked in case changes make it structurally unsound. The dead weight of bars and frames are usually OK unless using glass-glass panels. The majority of panels have Tedlar™ on the back rather than another layer of glass, so most modules and panels are sufficiently light enough for a structurally sound roof to support.

There is a need to check the roof is securely fixed to the walls of a building. Modern buildings tend to have galvanized steel wall-tie bars that fix roof trusses to walls, but older properties do not always have the roof timbers connected to the walls, they rest on the walls. Wall-roof ties might need to be retro fitted if not present. When fitting solar panels to a roof, it is common for wind uplift pressure to have a greater effect than the dead weight of the panels.

Tedlar™ Backing material used on many PV panels to reduce weight load

If there is any doubt about the ability of a roof to support the weight or uplift pressure then consult a structural engineer.

Check the roof/building in case changes make it structurally unsound

Must check that the roof is securely fixed to the walls of the building

If in doubt about the roof consult a structural engineer

UK AND INTERNATIONAL STANDARDS

Note: In the UK, further advice and training on fixing PV panels to roofs can be obtained from the National Federation of Roofing Contractors (NFRC) and the National Home Building Council (NHBC). NHBC publish a guide called NF30, 'Renewable energy systems of roofs on residential buildings' and this guide goes through some of the roofing loading calculations. In some countries, snow loading calculations are just as important as wind loading calculations.

Checking off-roof locations

Off-roof installations need preparation for the areas where they will be installed. The location for ground mounted systems needs checking for space, and the area the panels will be installed on their frameworks. It needs to be checked again for shadows and vegetation, such as fast growing grasses. The optimum angle of tilt and orientation also need reconfirming.

Pole mounted modules need very secure anchoring, as they are particularly liable to wind lift. The base needs to be prepared and installed before the panels can be mounted on the pole.

Check location for ground mounted system for space

Check for shadows and vegetation, e.g. fast growing grass

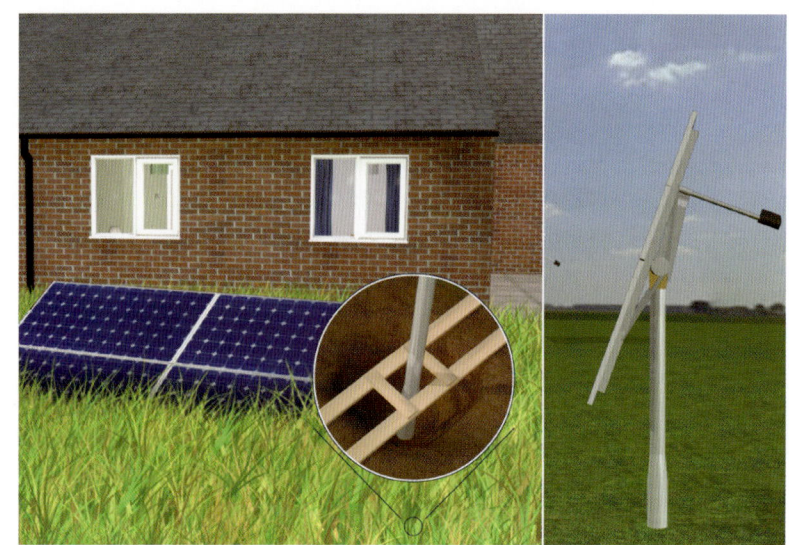

Pole mounted modules need secure anchoring - particularly liable to wind

ACTIVITY 45

For ground mounted PV systems, what extra considerations should be made as compared to building mounted PV?

PLANNING AND ACCESS

During the pre-installation phase the site can be inspected for the location of the major components and where they will be positioned.

At the same time a risk assessment can be made and any possible problems identified.

Work may need to be carried out in small and confined spaces, or at height, access may be restricted and you may need **RCD protection** – depending on where cables are positioned.

RCD protection
Ground fault protection. Full name is a residual current device

A risk assessment can be made to identify potential problems

HEALTH AND SAFETY

If installing a stand alone system check the battery store will comply with health and safety rules, as lead-acid batteries release Hydrogen during charging and the acid is highly corrosive. Provision should have been made to store batteries on a framework and insulated if necessary.

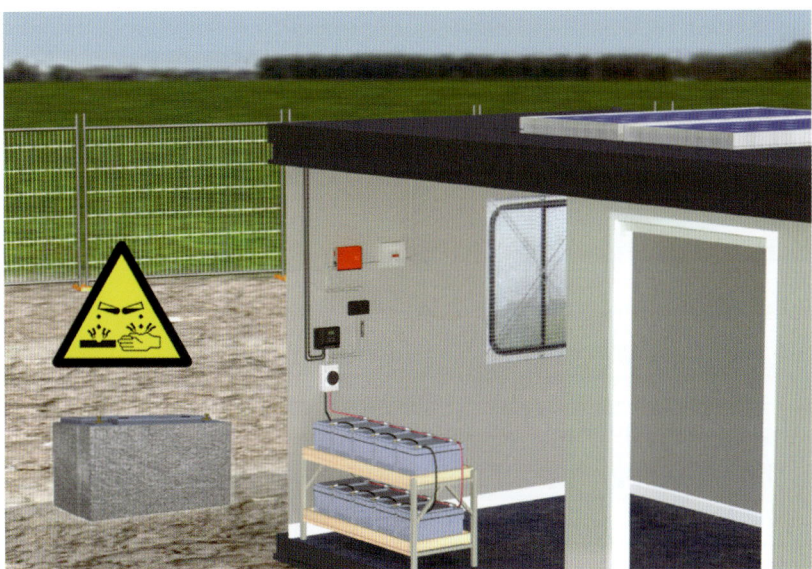

Battery store must comply with health & safety rules

Once the hazards have been identified, and locations assessed for risk, a plan of the work can be drawn up.

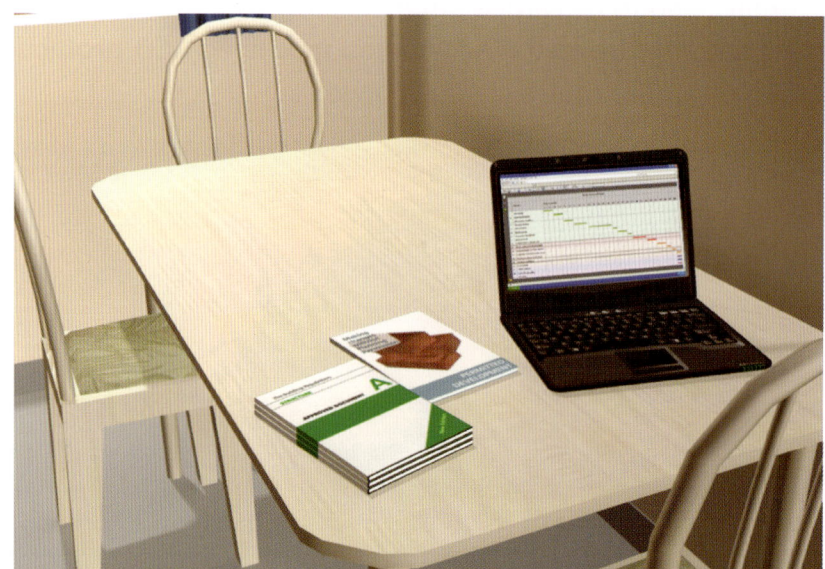

Once hazards are identified/risk assessments made, planning can commence

ACTIVITY 46

When covering a risk assessment on the installation process, should you just be concerned about your own and your colleagues, health and safety?

CHECKING THE MATERIALS

Checking against the plans

Pre-installation checks allow a final review of the plans and allow for any last minute changes that may need to be made:

- Is there anything that has been forgotten?
- Do the plans match the type of PV system being installed?
- Has the right equipment and cabling lengths been ordered for the positioning of the equipment?

For example, if the inverter is located in the loft and the generation meter in the garage, then it is often good practice to have an isolator with the inverter and the generation meter.

Pre-installation checks allow final review of plans and last minute changes

The inverter needs to be of an approved type and related to the feed-in tariff if appropriate. The inverter and meter displays can give a variety of information – will the customer be able to get the information they want from the equipment ordered? Extra monitoring information can include weather sensors; some or all of: the present power output in Watts, energy produced to date in kWh, CO_2 reduction since installation in Kg and the value of electricity generated in pounds sterling.

Inverter must be of an approved type

Optional connection to a home PC or laptop to download the display can be given, also comprehensive monitoring may be offered by the inverter manufacturer as an optional extra. This information can be shared and be public or private as the customer decides. These decisions should have been made earlier, but could be changed now if essential to do so.

ACTIVITY 47

You want to offer your customer extra reassurance that your company is reputable and reliable. What might you do to assist in this process?

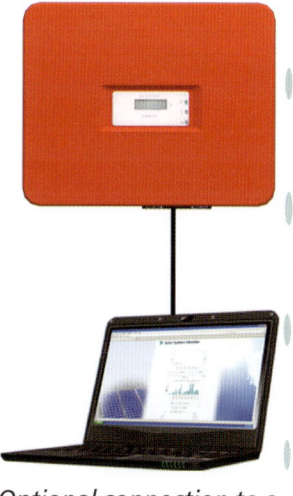

Optional connection to a laptop/PC to download display can be given

Checking materials

The major components ordered should arrive with datasheets, test certificates and guarantees. You need to check that they have been sent along with the items. Typical documentation includes:

- Details of PV module power output with tolerance limits
- Any test certificates
- Confirmation that inverters are suitable for grid connection in the UK, with factory certification and MPPT provided

Major components should come with datasheets, test certificates and guarantees

Visual inspection

Everything needs to be visually inspected. This is looking for surface damage as well as checking that what has been delivered is what was ordered. This includes the number of items – such as types of isolators, and the quality – such as size of cabling ordered.

Everything must be visually inspected

Testing for faults

All the components should be tested for faults before they are installed. This includes physical damage that may be visible, but also electrical damage can be tested. Using the datasheets, open circuit and short circuit readings can be tested and continuity tests can also be made. PV modules can be connected together as they will be in the array and checked for any out of range readings which are unexpected. Any out of range readings could be due to a wiring error or faulty module.

Any damaged items should be rejected and replaced.

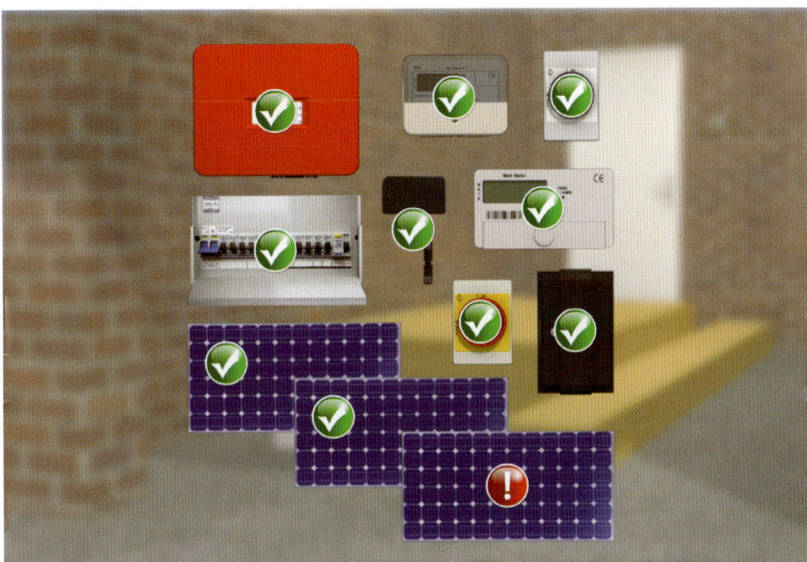

Test all electrical equipment for faults

E-LEARNING

Use the e-learning programme to learn more about checking materials.

ACTIVITY 48

Some installation companies have been known to only test the whole installation after all the PV panels are fitted, rather than each panel individually. What is the problem with this procedure?

Checking documentation

Once satisfied that all the equipment is present and of good quality, the documentation must be kept together. Access to installation instructions will be needed, and must be followed when installing the product. Various documents will need completing with readings taken during installation and commissioning, and finally given to the customer at handover.

The documentation must be kept together

CHECK YOUR KNOWLEDGE

1. **PV modules can be constructed in a number of ways. What type of combinations are shown here?**

Combination	Image
	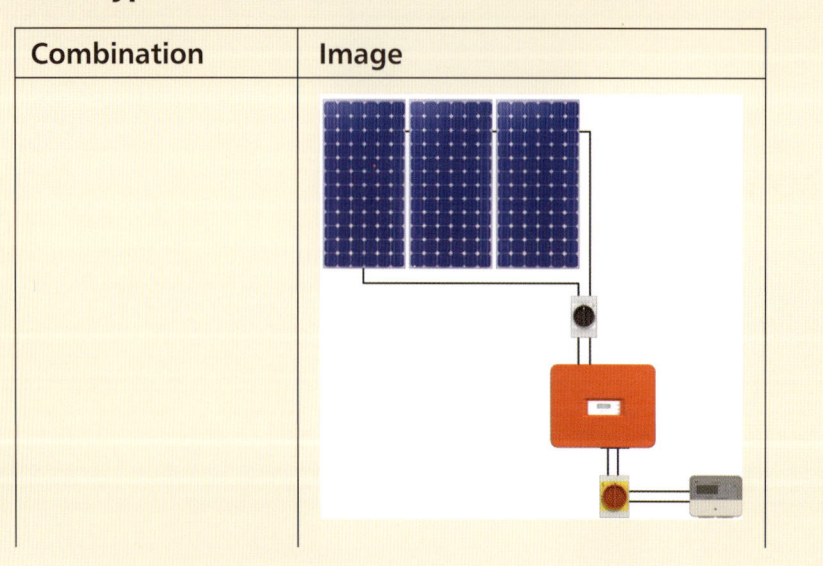

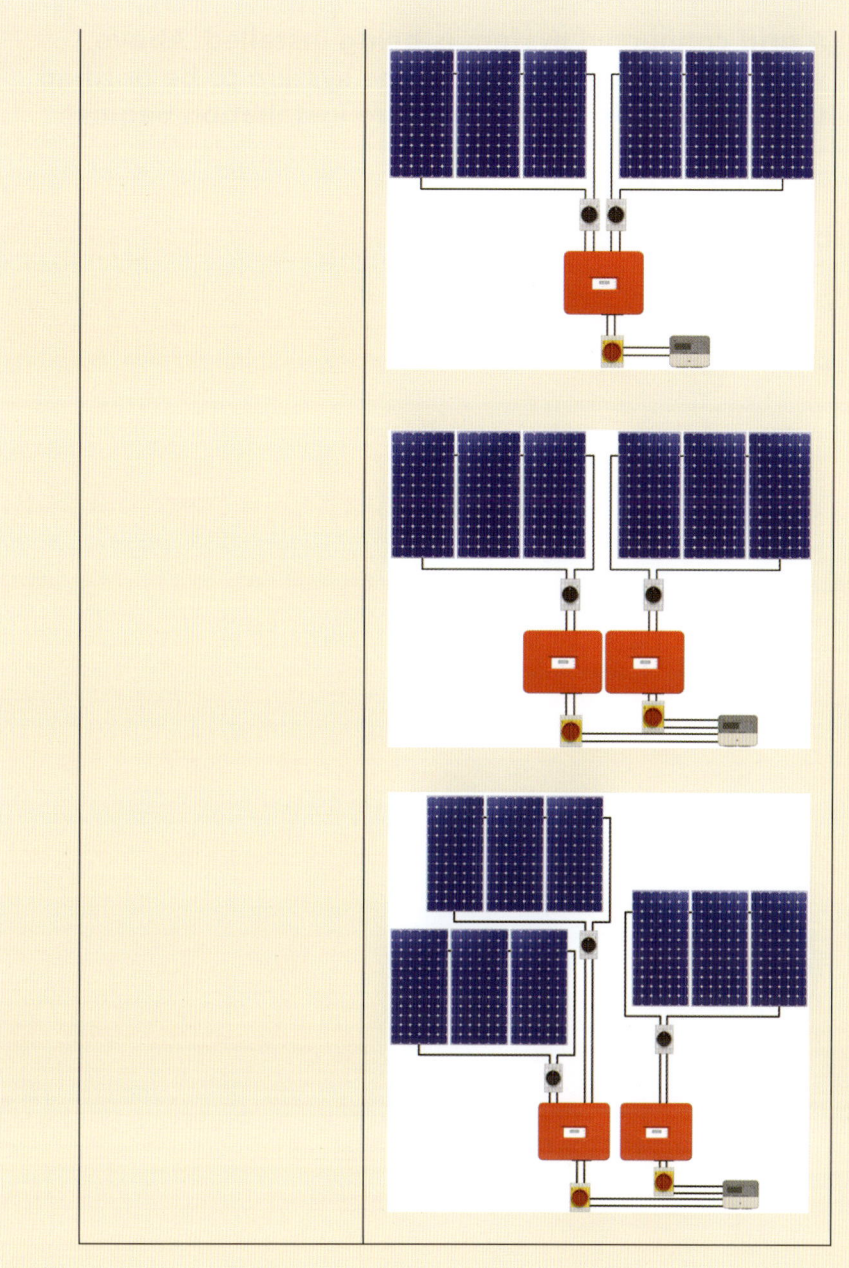

2. **When unpacking the PV modules they must be checked for physical damage and that they meet electrical outputs. What else can be checked?**

☐ a. Their colour

☐ b. Their type (crystalline, amorphous etc)

☐ c. Their weight against the documentation

☐ d. Their output when joined in an array

3. **A grid connected system is being installed. Above what value current requires the system to be notified to the local grid supplier before installation begins?**

☐ a. 12 amps

☐ b. 16 amps

☐ c. 20 amps

☐ d. 30 amps

Chapter 8

INSTALLATION

LEARNING OBJECTIVES

By the end of this chapter you will be able to:

- Identify the requirements for handling, moving and storing solar PV panels

- Describe the requirements for fixing 'on and in roof' systems and using secondary frame structures

- List the safety features that must be considered when installing solar PV systems

- Describe the requirements for fire and weatherproofing, ventilation and cables when installing solar PV systems

- Explain the particular importance of electrical safety when installing grid tied systems (electrical safety is important for all systems)

DC CABLES AND ISOLATORS

Installation of the DC cabling should be completed before the PV panels are installed. The cables should be connected into the DC isolating switch and secured where necessary along the length. They must be kept separate from any AC wiring in the house. Once in position they should be tested for continuity and insulation in case any damage occurred during installation.

Special labels should be fixed on connection boxes and isolators to show the cables are live during daylight, in readiness for when they are connected up to the modules.

Once the PV array has been connected, all testing or maintenance done on DC cabling must be done in the dark or with the PV array covered by opaque black covering.

The installation of the DC cabling should be complete first

Cabling between roof and garage

DC cabling needs to be kept as short as possible because power is lost in proportion to the square of the current. It also needs to be well insulated – to Class II standard.

DC cabling needs to be kept as short as possible

ACTIVITY 49

When selecting DC, what particular requirements should be considered? And where might you find out more information about the particular requirements to be considered?

Isolating switches

DC isolation switches and fuses must be capable of handling high DC current with contacts that will not arc. When installing they should always be set to the OFF position until testing starts.

DC cables should be connected to the DC isolator first, then the modules connected together forming the PV module string.

When installing DC isolation switches/ fuses they should always be off

Cabling on panels

PV array cables are often called array interconnects – and are resistant to UV light.

Interconnects are resistant to UV light

Cabling joining inverter

If more than one set of DC cables are being connected to the inverter take care to connect the pairs of cables correctly.

Cabling joiner inverter

If there are batteries

If batteries and a battery charge controller are being used, the DC cabling from the PV modules connects with the controller before connecting to an inverter or DC distribution board. The batteries link directly with the controller using DC cabling. There may be one or more DC isolators in this circuit. If the circuit has any DC appliances or lighting, they too will have DC cabling to outlet sockets.

The batteries link directly with the controller using DC cabling

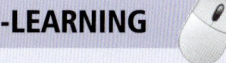

E-LEARNING

Use the e-learning programme to learn more about DC cables and isolators.

FRAMEWORKS FOR PV MODULES

Building access scaffolding and procedures for fitting the PV modules have associated hazards that must be assessed before installation takes place, including working at height, manual handling, risk of electrocution, risks due to falling objects and risk from working in confined spaces.

Building access scaffolding has associated hazards

PV modules must have sufficient ventilation. Therefore roof systems often have standard frameworks for support and to keep a gap for ventilation behind the modules. Integrated PV materials need sufficient ventilation in the attic space. Most roof mounting systems will be attached to rafters in the roof structure. Sometimes counter-battens can be added to increase ventilation. In all cases follow the manufacturer's instructions.

PV modules must have sufficient ventilation

ACTIVITY 50

For an in-roof PV installation, how would you design the system so as to allow for adequate ventilation?

Pitched roofs – roof hooks

Roof hooks A system for attaching framework to a roof commonly used with tiled roofs

On-roof PV systems can be mounted in portrait or landscape orientation. Thin film modules are typically mounted so the strips are vertical. By using standard modules, standard frames can also be used. Only a few roof tiles or slates need to be removed and then replaced to fix the frame to the rafters. Generally, **roof hooks** work best with tiled roofs.

On-roof PV systems can be mounted in portrait/landscape orientation

The roof is carefully measured to fit the modules to the correct area. The tiles are removed and the roof hook is screwed to the rafter then the tile replaced. Tiles may need to be notched slightly. The PV rails are then bolted to the exposed part of the roof hook.

The roof must be carefully measured to fit the modules

The PV rails are bolted to the exposed part of the roof hook

Pitched roofs – stand-offs

Stand-offs are secured to the rafters, and work best on slated roofs. Stand-offs have a square plate and tube which is sealed into the roof to make it weatherproof. A plate is then welded to the top of the tube, to which the array rails can be bolted.

> **Stand-offs** A system for attaching framework to a roof, commonly used with slate roofs

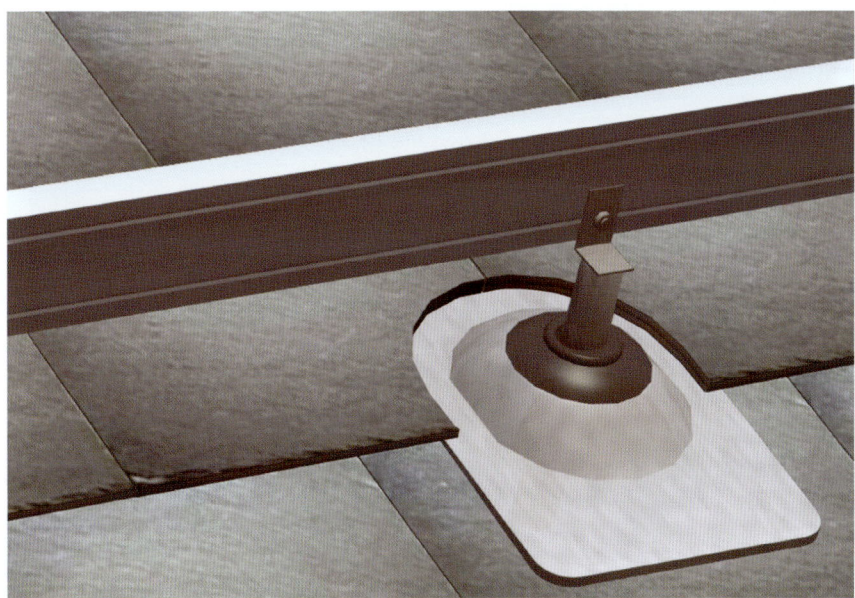

Stand-offs work best on slated roofs

Integrated roof options

Integrated PV modules can be of two main types. One uses PV modules, usually of laminates set into a frame attached to the rafters. The framework is built first then the modules attached and the flashing added to complete the weatherproof seal.

The second option uses PV tiles or slates made from amorphous or crystalline PV cells. The surface area of PV material tends to be larger than modules because there are more gaps between the cells. Aesthetically the building will maintain its original look and this option is frequently required for listed buildings.

The framework is usually built first

The modules are then attached and the flashing is then added

ACTIVITY 51

Try to name six or more types of pitched roof types and which of these types would probably not be suitable for fitting PV?

Flat roof frameworks

Flat roofs need substantial galvanized steel frames. They can have optimum orientation and tilt. This can be confirmed just before fitting them. Fixing the frames to the roof is not always sufficient to stop them blowing away so the frames are weighted down with dead weights – typically concrete blocks. If the frames are attached to the building, they must be very securely fixed, often 600mm or more into the brickwork below.

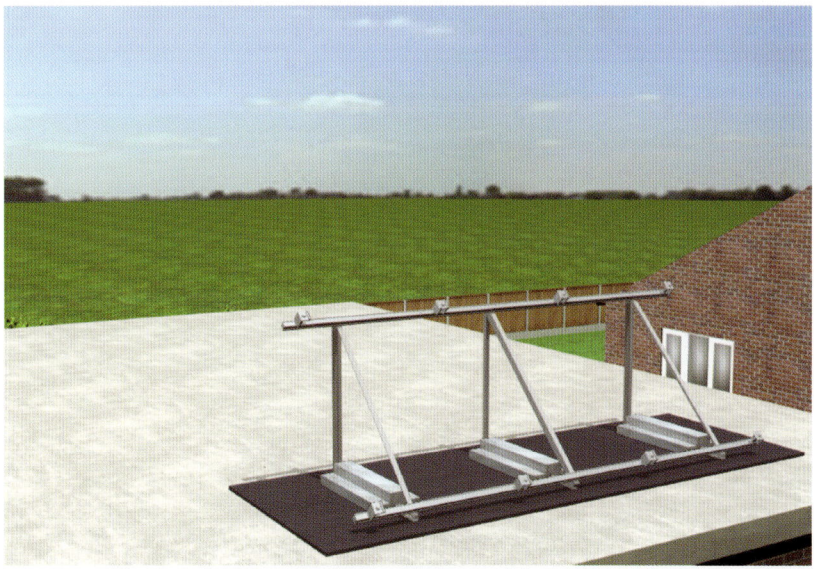

Flat roofs need substantial galvanized steel frames

Flat roofs can have optimum orientation and tilt

Ground and pole frameworks

Ground mounted options are similar to flat roof frameworks that can be orientated at the optimum angle and tilt. Ground mounted frames may also be fitted with solar trackers. The fixings to the ground can be deeper and more secure and dead weights are unlikely to be needed.

Pole mounted options must have the concrete, secure base and anchor points built and finished before the PV panels can be fitted.

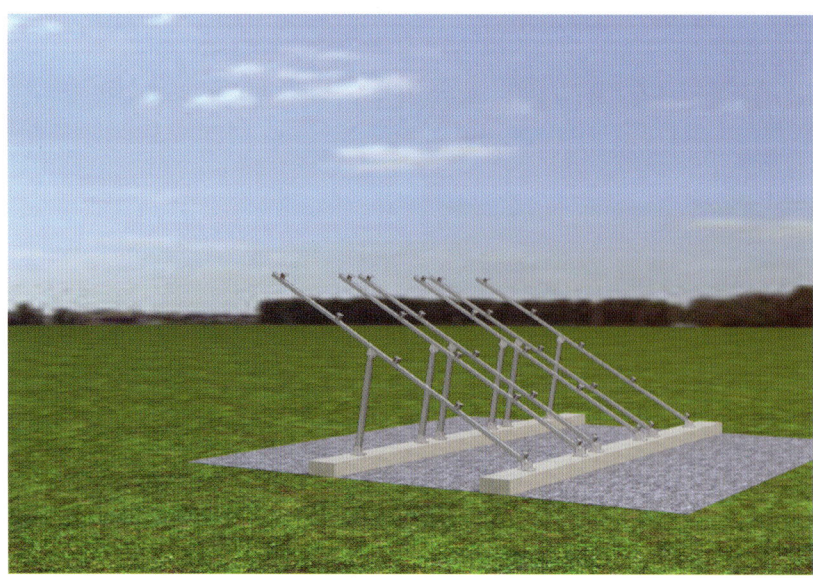

Ground mounted frames can be fitted with solar trackers

Ground mounted options can be orientated at the optimum angle and tilt

Pole mounted options must have concrete, secure base and anchor points

ACTIVITY 52

Some suppliers are offering solar mounting frames of very light weight that are mounted on top of a flat roof and use wind pressure to hold down the PV panel array. To explain further, whatever direction the wind blows, the array and its frame push down onto the roof structure. What guarantees would you probably want before fitting such a roofing system?

INSTALLING PV MODULES

The PV modules may be cleaned with glass cleaner that repels water and dirt before they are installed in their positions.

Great care needs to be taken when manoeuvring the panels into place. Not only are they fragile, they may also be caught by the wind.

HEALTH AND SAFETY

They will also be generating electricity and therefore all the precautions of electricity at work apply. A panel is often around 24 volts and once joined into an array the voltages can add up to give quite a powerful jolt!

PV modules may be cleaned with glass cleaner

As the panels are installed they need to be connected together. The connectors are modular and can only be linked together in one way when building the strings. They must be secured to the framework or roof as appropriate. Connect the panels in series or parallel according to plan, following the manufacturer's instructions. The cabling will be UV resistant and weatherproof. They are underneath the modules and therefore not easy to access once the panels are installed.

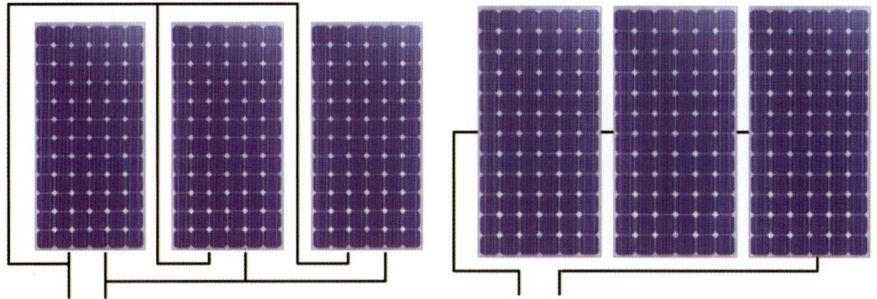

The panels need to be connected together when being installed

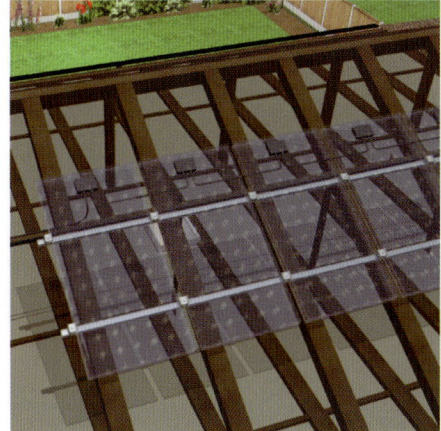

*Panels must be secured to the framework
or roof as appropriate*

Recording details

As the panels are installed record the serial number and location of each panel and any other details that need to be recorded not already taken, such as the **open circuit voltage** and short circuit current.

Open circuit voltage (V$_{oc}$) Voltage across an illuminated PV cell or module when there is no current flowing. It is the maximum possible voltage

Product Code: 123456
Seriel No: 123-456-789-00
Location: *ROOF1*

Record the serial number and location of each panel

INSTALLING AN INVERTER

The inverter needs plenty of space around it as it can get quite hot. Make sure the manufacturer's instructions are followed. The display is given in AC output, not DC input.

In the UK the inverter size is often about 20 per cent smaller than the array size, as it operates below maximum rating for most of the time. Inverters can operate above maximum rating for short periods of time. Some inverters can be very large and heavy, so make sure the wall can take the weight.

Ensure that the inverter has plenty of space as they get quite hot

ACTIVITY 53

Would you follow similar inverter sizing procedure and fit an inverter about 20 per cent smaller than PV array peak power capacity in southern Europe?

Start up and shut down procedures

When connecting an inverter, the DC and AC isolators must be turned to OFF. During operation always disconnect the AC output before disconnecting the DC side while PV panels are producing electricity.

Always isolate the DC output from the panels before disconnecting or connecting any DC connections.

To connect the inverter the DC and AC isolators must be turned off

AC WIRING AND ISOLATORS

Regulations

UK AND INTERNATIONAL STANDARDS

The existing AC system within the house will have been checked before work started. When a grid connected system is installed the changes will be made at the distribution board and the PV system, rather than the rest of the house – assuming the existing wiring was OK. All grid linked solar PV systems must conform to either G83 part 1 or G59 part 1 – depending on whether the current produced is under or over 16 amps. Otherwise the system should conform to BS 7671 as usual.

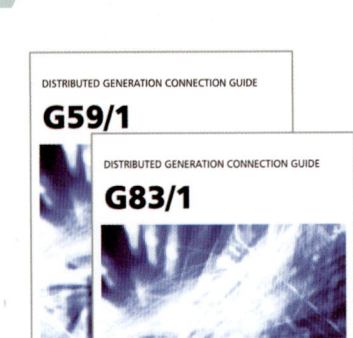

All grid linked solar PV systems must conform to G83/G59 Part 1

The existing house AC system will have been checked prior to starting work

ACTIVITY 54

In the above, we list the British PV grid connection protocols, G83/1 and G59/1. What are the benefits of having national grid connection protocols and what would happen if they were missing?

AC isolators

Between the inverter and the distribution board there must be an AC isolator and the generation meter. Depending on the positioning of equipment this will be a double pole isolator placed near the inverter. There may be an additional isolator if the distance between the inverter and consumer unit is long, such as between floors of the building. RCD protection may also be included.

There must be an AC isolator/generation meter between the inverter and distribution board

INSTALLING A GENERATION METER

The generation meter measures the AC electricity generated from the PV panels and is situated between the inverter and the consumer unit. It must be of a type approved by the local network grid supplier and Ofgem. The feed-in tariff is measured on this meter. Installation may be carried out by the local network supplier.

The generation meter must be approved by local network grid supplier and Ofgem

ACTIVITY 55

In the UK, Ofgem is the regulator responsible for heating and electricity. All nations must have a system for regulating electrical systems and standards. What would a typical national regulatory framework include?

DISTRIBUTION BOARDS

The distribution board, or consumer unit, is connected to the electric supply from both the local grid network and the PV generated electricity passed through a double pole isolating switch, then to the domestic circuits. Various arrangements are possible. Shown here is a typical unit. One diagram is a distribution board with RCD protection on some circuits, and one with an auxiliary consumer unit positioned elsewhere on the premises.

Distribution board

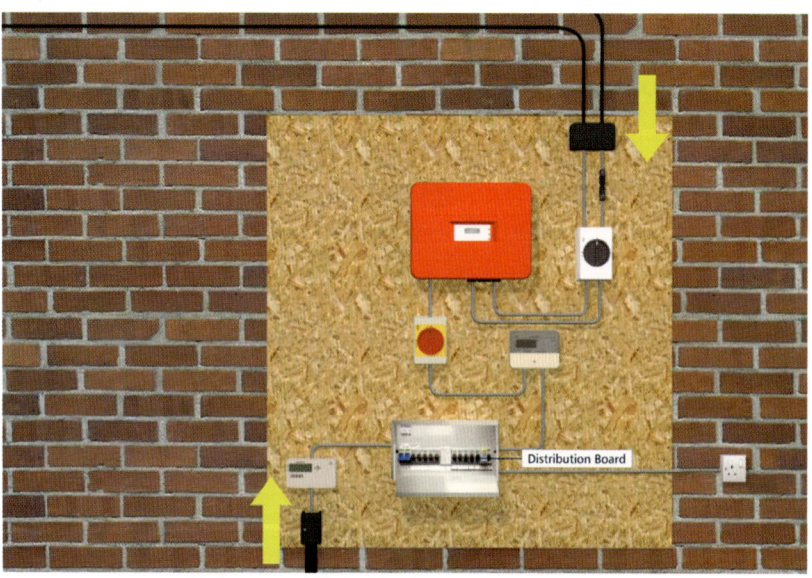

Auxiliary consumer unit positioned elsewhere on the premises

Consumer unit with RCD protection:

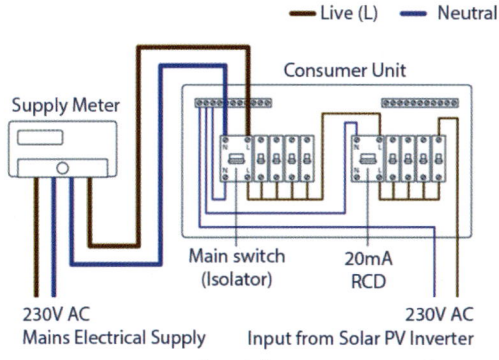

Consumer unit with RCD protection

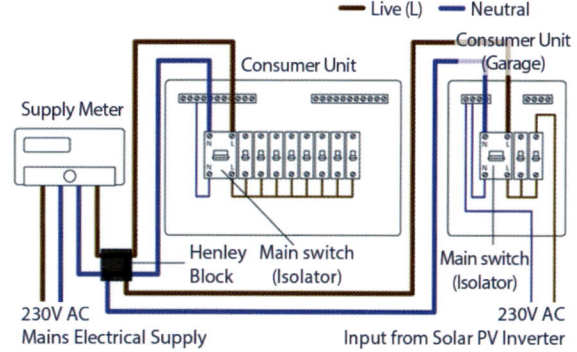

Supply with two separate consumer units, supply to one unit only

ACTIVITY 56

Because there is a consumer unit (aka distribution board) with plenty of MCB and RCD protection, there is no need for fuses in the PV system. Is this statement true or false? Please discuss this statement further.

INSTALLING BATTERY SYSTEMS

Systems with batteries are usually stand alone systems. The battery store must be well ventilated and racking should be installed first. This is to keep the batteries off the floor and allow insulation underneath as well as round the sides of the batteries.

HEALTH AND SAFETY

Great care must be taken when installing the batteries as they are full of corrosive acid and can be very heavy. Never put insulation on top of batteries as it impedes the flow of escaping gases from lead-acid batteries.

The batteries will be connected in series or in parallel using battery connectors and DC cables. A DC isolator will be needed to shut off the battery supply before it reaches the load and charge controller.

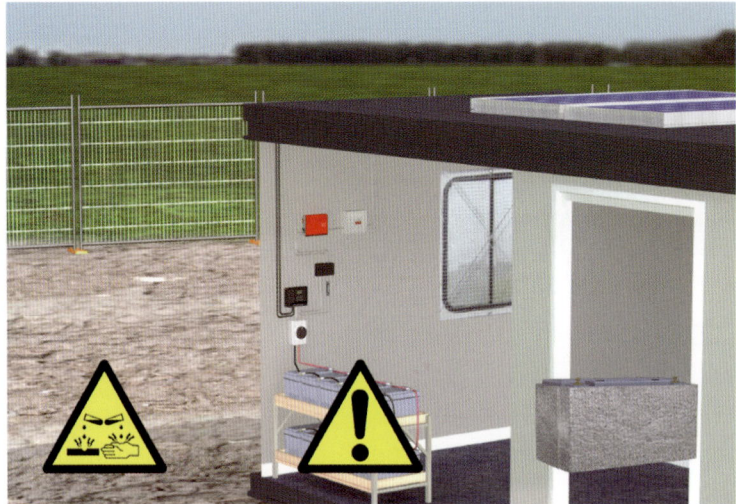

Systems using batteries are usually stand alone systems

INSTALLING GRID TIED SYSTEMS

Single phase grid connections

The grid operates with a three phase alternating current, which means the cables have five wires rather than three: one neutral, one earth and three live. Each of the live wires has the electrons moving at slightly different times, giving a phase interval between them. Domestic premises use only one phase and are therefore connected to one of the live wires from the grid. In any road, houses are connected to different live wires and this spreads demand.

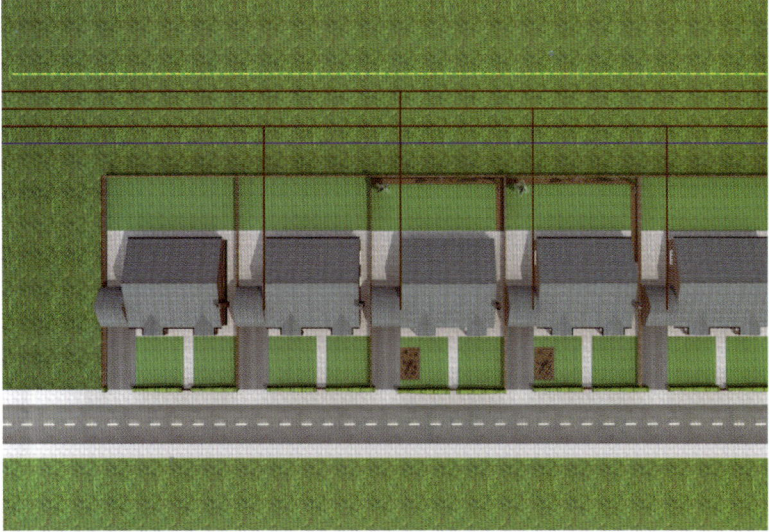

Cables have five wires: one neutral, one earth and three live

When connected to the grid, Small Scale Embedded Generators or **SSEG**s are also linked to one of the live wires when feeding back to the grid, provided they do not produce more than 16 amps and come under regulation G83 part 1. The DNO needs to check that other

SSEG Small Scale Embedded Generator

SSEGs in the area are not also feeding back into the same live wire, as the DNO needs the phases to be balanced.

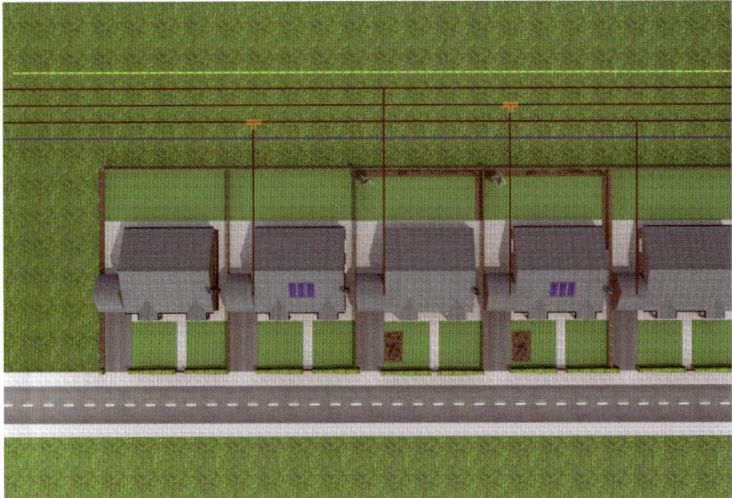

Small Scale Embedded Generators (SSEG) also linked to live wires

ACTIVITY 57

As well as solar PV, name some other potential sources of SSEGs.

Three phase grid connections

Commercial and industrial properties usually have grid connections to all three phase live wires and then distribute single phase within the building.

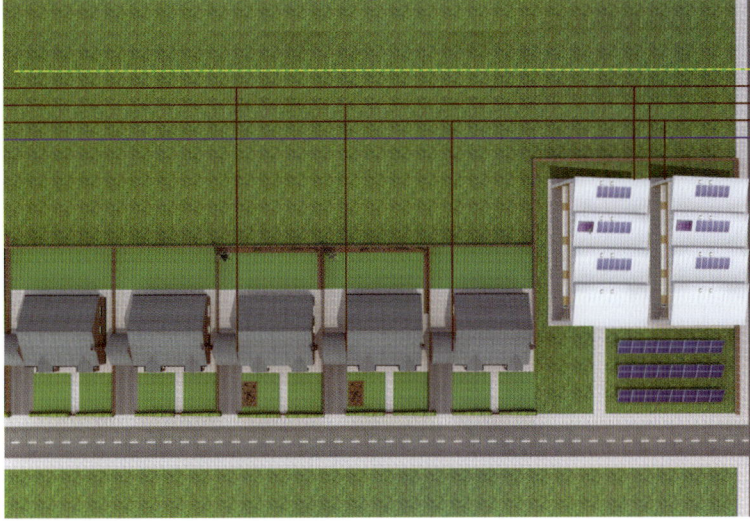

Three phase grid connections

If the PV installation will produce more than 16 amps the DNO may prefer to have a three phase feed back into the grid to keep the current between the phases balanced.

It is important to keep the current between phases balanced

CHECK YOUR KNOWLEDGE

1. What two items must have good ventilation?

☐ a. PV modules

☐ b. Inverter

☐ c. DC isolator

☐ d. Generation meter

☐ e. Import meter

☐ f. Distribution board

☐ g. DC cables

2. What is the most dangerous natural hazard when manoeuvring PV panels to their position on a roof?

☐ a. Sunshine

☐ b. Rain

☐ c. Wind

☐ d. Snow

Chapter 9

PV PROTECTION

LEARNING OBJECTIVES

By the end of this chapter you will be able to:

- Describe the effects of system exposure to excessive voltage, current or frequency events

- Describe techniques and components to protect the DC circuit from excessive voltage and current events

- Describe techniques and components to protect the AC circuit from excessive voltage and current events

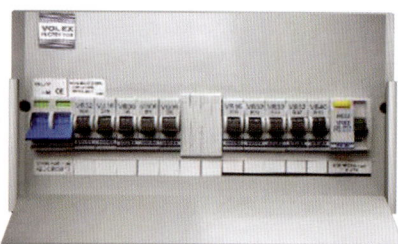

Bungalow with solar PV panels installed

Isolation switch

Fuse box

EXTREME ELECTRICAL EVENTS

Over and under the expected

Low voltage systems can cause major problems as it is possible to get excessively high bursts of current for very short periods. For people this can cause shock or death, but also cables would run hot and create a fire hazard or burn someone. These short circuits can also damage equipment. They can reach 1000 plus amps.

Unexpected high voltage can also hurt anyone working on equipment and damage equipment. Modern maintained AC circuits in domestic settings have protection built into their existing power supply already through the circuit breakers and fuses; however excessive surges can still cause shocks.

Low voltage systems can cause major problems

The ultimate unexpected DC current event is lightning, which is very over voltage and over current! Except it is not so unexpected. Lightning happens fairly frequently in the UK and metal frames on exposed roofs or ground can attract lightning unless precautions are taken. Lightning strikes nearby can be dangerous to the PV system and cause great damage to the panels and internal equipment. Always follow the manufacturer's instructions relating to lightning protection.

Lightning is very over voltage and over current

ACTIVITY 58

The risk of a direct lightning strike on a typical roof mounted solar PV installation is generally considered to be low, but where the building in question is either tall or in an exposed open area, making it the highest structure in that area, there may be a greater risk. What actions should you take if there is a greater risk? Hint, 'Photovoltaics in Buildings, Guide to the Installation of PV Systems' should assist here.

Islanding

It is important that if the main grid goes down, the PV system also shuts down so that the grid can be maintained without electricians being electrocuted.

Therefore, grid tied inverters normally shut off if they cannot detect the presence of the grid, so if the grid suffers a power failure then the PV feed-ins to the grid should be cut off too. But if loaded circuits nearby resonate at or near the grid frequency (50 Hz) then the cut-off device can be fooled into thinking the grid is still active. This process is called **islanding**.

Inverters for use with the grid have anti-islanding protection by injecting small pulses at out of phase frequencies that then cancel any stray resonance effects when the grid shuts down, thus giving an anti-islanding effect, cutting off the PV supply and preventing accidents to anyone who may be working on the grid trying to carry out repairs.

Islanding Situation that occurs if a grid connected PV system continues to supply electricity to the grid during a power cut. This can cause electrocution to technicians working on the grid, but can also create an explosion hazard when power is restored to the grid

If the main grid goes down, the PV system must shut down

ACTIVITY 59

Islanding inverters feature automatic disconnection from the grid in the event of grid power failure. What other safety and/or grid connection features will an inverter have and what are the two main points to consider when mounting an inverter?

DC PROTECTION

There are various ways of mitigating for excessive voltage and current in the DC circuits of PV systems.

Many ways to mitigate for excessive voltage and current

Isolation switches

Isolation switches must be placed between the inverter and the array – and if there are long cables, at both ends of those cables. Always start with the isolator in the disconnected OFF position, and only turn ON when you know it's OK.

A DC isolation switch is needed for each string, which will allow the open circuit voltage and short circuit current to be safely measured.

Fuses are used in the circuit too, although for higher voltages than 12V or 24V DC systems, special DC fuses are needed.

Isolator must be off when starting, only turn on when safe

Batteries

A DC protection fuse is built into the controller and is normally sufficient for basic circuit protection. Most small systems should have the negative terminal on the battery adequately earthed. An earthing rod can be used if no suitable earth is available. Larger battery systems should also have the positive battery terminal connected to a fuse; when connecting them up ensure all current from batteries passes through that terminal.

A DC protection fuse is built into the controller

ACTIVITY 60

Are batteries always required in grid disconnected PV circuits?

Earthing and bonding

Bonding will only typically be required if the modules, DC cables and connectors are not Class II and the frames are considered to be an extraneous conductive part, as defined in the Wiring Regulations.

Bonding of the PV module frames can reduce the risk of electric shock to anyone coming into contact with the array. Such bonding of module frames also provides some protection from lightning strikes. However, bonding of the array frames is not always required. Bonding of the PV modules would NOT be required as long as ALL three of the following conditions exist:

1. The modules are Class II. Many PV modules meet this requirement.
2. The cables, connectors, junction boxes are Class II. Many manufacturers' cables, connectors and DC isolators meet this requirement.

3. The inverter has an isolating transformer between the DC and AC parts. Many manufacturers' inverters meet this requirement as they have the necessary Galvanic Isolation, as required by G83/1–1.

If a PV system is not earthed it must be Class II insulation, i.e. double-insulate. PV modules are normally Class II insulated when supplied.

Lightning protection

The risk of a direct lightning strike on a typical roof mounted solar PV installation is generally considered to be low, but where the building in question is either tall or in an exposed open area, making it the highest structure in that area, there may be a greater risk. In such a case, advice should be sought from a lightning protection specialist and, where necessary, a lightning protection system installed conforming to BS 6651: Code of Practice for Protection of Structures against Lightning.

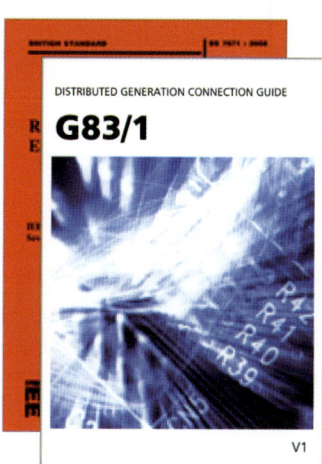

Distribution generation connection guide - G83/1

The risk of lightning strike on roof mounted panels is low

ACTIVITY 61

BS 6651: Code of Practice for Protection of Structures against Lightning has been replaced by BS EN 62305. Use the BSI (**British Standards Institution**) to look up what is the new Standards title, how many parts are in the series and how many lightning strikes hit the UK every decade.

British Standards Institution (BSI) sets quality standards and standard dimensions for equipment and materials. All British Standards start with the letters BS followed by a number

Positioning of equipment

If not at risk from a direct lightning strike, a solar PV system can also be damaged from induced surges in the power supply, particularly where this is via overhead power lines. Power surges due to lightning can occur in either the DC or AC parts of the system. To protect against surges the DC cable runs should be kept as short as possible, and the positive and negative cables from a string or main DC supply cables should be bundled together. Long runs of DC cable, for example over 50m, should be installed in earthed metal trunking or conduits to shield them from induced surges.

Panels can be damaged from induced surges in power supply

E-LEARNING

Use the e-learning programme to learn more about DC protection.

ACTIVITY 62

To protect against surges the DC cable runs should be kept as short as possible and the positive and negative cables from a string or main DC supply cables should be bundled together. Long runs of DC cable (e.g. over 50m) should be installed in earthed metal trunking or conduits to shield them from induced surges. Why is the above recommended and what extra measures should be taken with the trunking? If the DC cable is not fitted in trunking, what type of DC cable should be used?

AC PROTECTION

Before the PV system is installed, the existing AC circuits will have been checked and therefore have normal protection.

When the PV system is installed, the AC isolation switch needs to be placed where the inverter output is connected to the building power supply. The local network supplier may require the isolating switch to be located on the consumer unit.

Grid tied systems should also have a DC ground fault interrupter, or RCD protection, fitted. This is good practice as they detect current leakage into the ground from an ungrounded source and turn off the system. However, nuisance tripping might occur so always follow manufacturers' advice.

AC circuits should be checked before installation of PV system

Always follow the manufacturer's instructions

SYSTEM PROTECTION

PV system protection

Protection is required in any system to ensure it shuts down safely in event of a short circuit. PV systems must be able to turn off the electricity generated from the PV panels. However, PV systems connected to the grid have two sources of electricity: DC panels and the grid, or DC panels and DC batteries; or, in some cases, DC panels, DC batteries and the grid.

Start up and shut down procedure needs to follow the order of always disconnecting the AC output from an inverter BEFORE disconnecting the DC side whilst PV panels are under load. Always isolate DC output from panels before disconnecting or connecting any DC connections.

Protection is required to ensure safe shut down in event of a short circuit

CHECK YOUR KNOWLEDGE

1. **What part of this grid connected PV system will protect it against frequency events that could stop the system automatically cutting off from the grid when a grid power failure occurs? Circle the correct component.**

2. **What can be the results of a PV short circuit without any electrical protection?**

☐ a. Overheated cables

☐ b. PV panels damaged

☐ c. Fire

☐ d. Electric shock to anyone working on circuit

☐ e. Batteries charged more quickly (if present)

☐ f. Greater supply to grid feed-in tariff

☐ g. Damage to electrical components on the short circuit

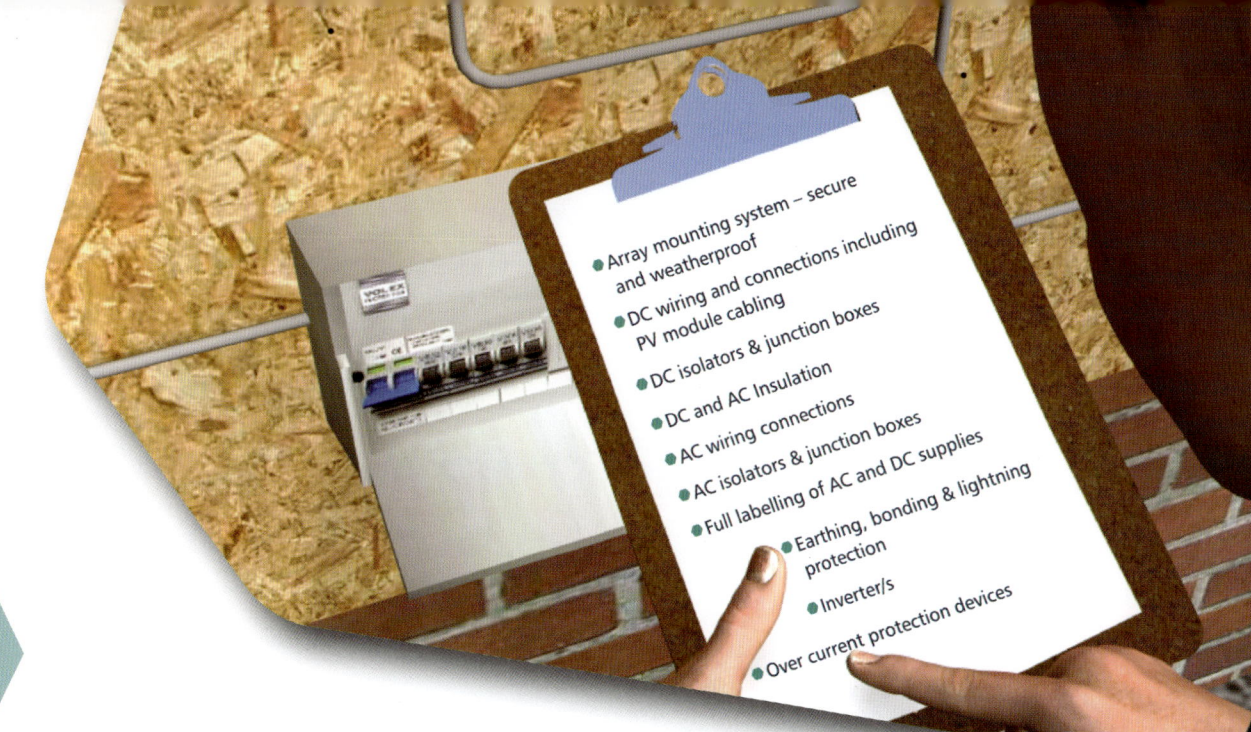

- Array mounting system – secure and weatherproof
- DC wiring and connections including PV module cabling
- DC isolators & junction boxes
- DC and AC Insulation
- AC wiring connections
- AC isolators & junction boxes
- Full labelling of AC and DC supplies
- Earthing, bonding & lightning protection
- Inverter/s
- Over current protection devices

Chapter 10

COMMISSIONING

LEARNING OBJECTIVES

By the end of this chapter you will be able to:

- List the requirements to commission a solar PV system

- Identify the conditions required to commission a solar PV system

- List the requirements to test and commission the AC circuit within a solar PV system

- List the requirements to test and commission the DC circuit within a solar PV system

- Describe the relevant regulatory requirements with respect to the manufacturer's requirements

House with solar PV panels installed

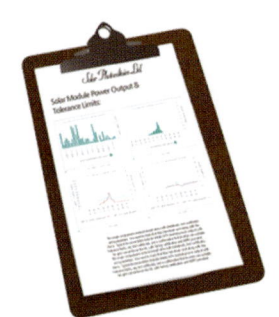

Requirements

PREPARATION FOR COMMISSIONING

What do you need to know?

The inspection and testing involved during commissioning requires a competent person – usually a qualified electrician – and that person can be identified at the start of the whole process.

The commissioning needs to be carried out matching the installed equipment to the design specified earlier. This may include specified makes of equipment as well as appropriate size and positioning of items in the house.

The conditions for carrying out commissioning need to have some daylight – a PV system cannot be tested properly if the weather is very overcast or it is dark, such as after sunset. Sunny days are best as the outputs from the system are to be matched against expectations. The PV system must also be tested under the Electricity at Work regulations, as for all new electrical systems.

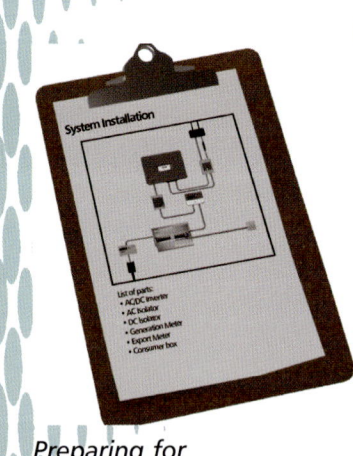

Preparing for commissioning

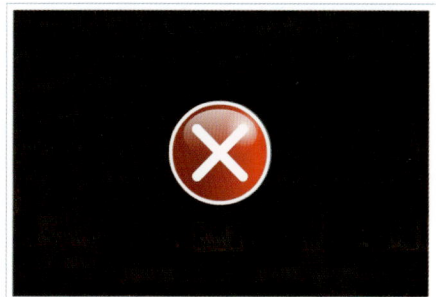

Commissioning must be carried out in daylight

PV system must be tested under the Electricity at Work Regulations

WHAT DOES COMMISSIONING INCLUDE?

Commissioning checks the integrity of newly installed systems and identifies any faults that may have occurred before the system is energized. Various visual checks are made first. The PV modules need to be checked, and check also that the array mounting system is secure and weatherproof. The DC circuits and AC circuits need checking and testing. All labels must be present on the wiring and equipment. The safety measures also need checking, to see that they work as intended, and the inverter needs to be checked to see if it's in the right place, with sufficient ventilation around it. Schedules of tests and checks are available on the Internet.

Commissioning checks the integrity of the system and identifies faults before it is energized.

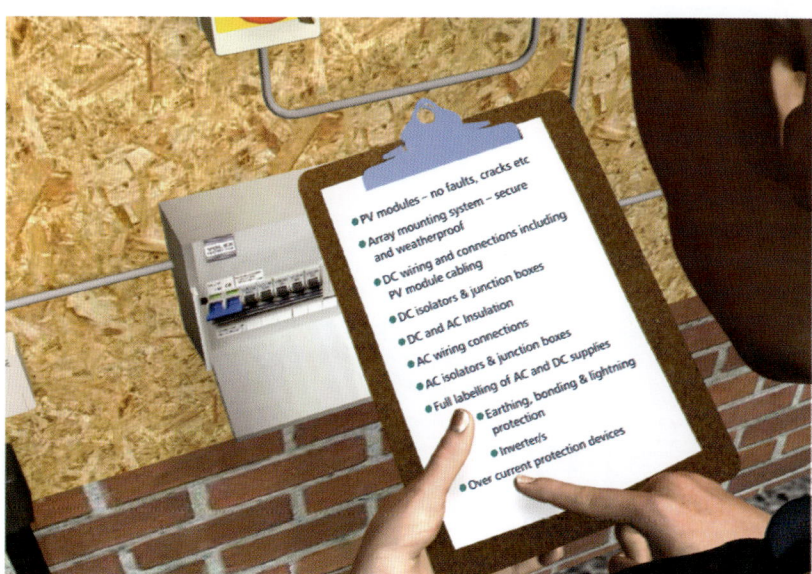

Commissioning includes the checks listed in this image

Some tests could be fatal if carried out incorrectly/at height

Visual checks need recording on the schedule, and if anything is not right it should be sorted out before the system becomes active.

When the tests are made according to the schedule, remember that some tests could be fatal if carried out incorrectly or at height!

ACTIVITY 63

What specialist commissioning tooling will be required?

Importance of records

Throughout the commissioning process, test readings and measurements are taken and it is important to record them as they are taken. Serial numbers also need recording, although it is best to note them down during pre-installation when checking the equipment.

The readings will be needed to complete the documentation which is handed over once the system is commissioned. All subsequent checks will be made against these initial readings.

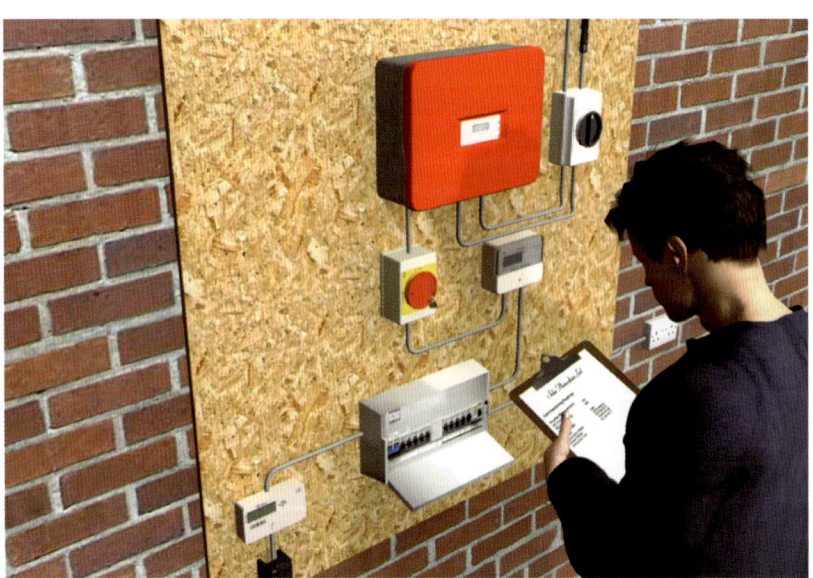

Record test readings and measurements taken during the commissioning process

Specialist commissioning

Most domestic installations are simple and straightforward, but occasionally specialist commissioning is required when installations are not in this category. Examples would include large domestic and small industrial installations, or where extra planning permissions or building regulations have been required, or where more than 16 amps are supplied back into the local grid.

Additional arrangements would be required to carry out specialist commissioning. If in doubt seek further assistance.

Most domestic installations are simple and straightforward

COMMISSIONING THE ARRAY

Module testing

The individual PV modules making up the strings and array require a variety of tests and measurements to be recorded. Although some readings can be taken during commissioning, others required to complete the paperwork are best taken during the installation process. They can be difficult to take once equipment is installed.

All the module readings need to be compared with those issued by the manufacturer. The readings will need to be adjusted for temperature and **irradiance**.

Irradiance
Instantaneous solar power received on a given surface, measured in W/m²

The individual PV modules require a number of tests and measurements

Array testing

solar irradiance Solar radiation at a location measured in kWh/m²/ annum. Irradiance is instant solar power in W/m². Solar radiation is also sometimes called insolation

Two tests carried out on the array include the open circuit voltage and short circuit current. When the array is tested and checked the **solar irradiance** levels must also be taken, preferably when it is close to the maximum, such as in the late morning or early afternoon on a sunny day. The temperature also needs to be taken into account to accurately measure the output from the array. The expected output can then be calculated and compared to the actual output. If there are discrepancies the wiring probably needs to be checked first.

Freestanding arrays will also need to be checked during commissioning to see that the plane of the array is correct for its location.

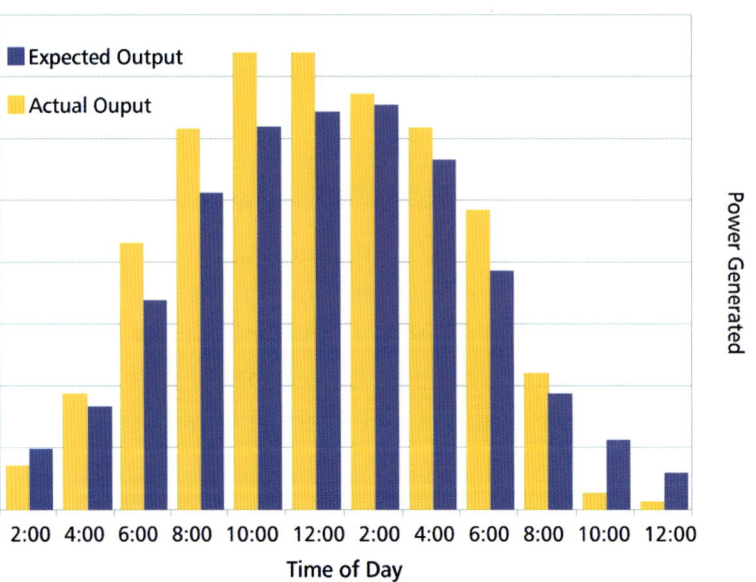

Graph showing the power generated against the time of day

COMMISSIONING THE INVERTER

The inverter will give information about the output of the PV system. It will lose about 5 per cent to 10 per cent of the generated power when converting from DC to AC current and any fault needs to be identified and rectified as soon as possible.

With grid connected PV systems with a feed-in tariff the local DNO must be notified before the link to the grid can be connected.

The inverter will lose 5–10 per cent of generated power when converting from DC to AC

ACTIVITY 64

What will most inverters display? Is this inverter acquired information enough for commissioning and handover?

DOCUMENTATION

The documentation must be completed with measurements and readings taken during installation and commissioning. This documentation forms part of the handover for the customer, to be kept by the equipment – usually the inverter – so it is available for anyone called in to maintain or test the PV system. The DNO must also be notified and documents completed and sent to them.

Documentation must be complete with measurements and readings

Customer documentation

As a minimum, customers need to have the following documentation:

- Electrical installation certificate
- Schedule of inspections
- Schedule of test results
- PV array test report
- PV installation check list

Customer documentation

ACTIVITY 65

As well as the above minimum documentation, what other documentation should you provide for the customer?

DNO documentation

Grid connected PV systems under 16 amps per phase, under G83/ Part One, must inform the DNO that a small scale embedded generator has been installed onto their network. Documentation confirming this must be sent to the DNO at or shortly after commissioning.

Installations producing more than 16 amps per phase must have the advice from the DNO at the design stage. DNO documentation for these installations can differ.

UK AND INTERNATIONAL STANDARDS

Note: Always follow the connection and commissioning requirements of your local or national utility. G83/1 is provided here as an example of a British grid connection protocol. All protocols are subject to change and dependent on local requirements.

G83/Part One

CHECK YOUR KNOWLEDGE

1. **PV systems producing more than 16A can be connected to the local grid and confirmation sent to the DNO to comply with G83/Part 1 at commissioning. Is this statement TRUE or FALSE?**

   ☐ a. True

   ☐ b. False

2. **What part of this grid connected PV system will give you the measurements used to calculate the efficiency of the array?**

3. **Who will always have the particular competency required to carry out the commissioning of a new solar PV installation?**

   ☐ a. Installer

   ☐ b. DNO representative

   ☐ c. Structural engineer

   ☐ d. Qualified electrician

Chapter 11

HANDOVER AND MAINTENANCE

LEARNING OBJECTIVES

By the end of this chapter you will be able to:

- Describe the handover procedure

- List the required information to be given and demonstrated to the client

- Identify the performance monitoring required for the system

- List the expected maintenance required for the system

Handing over to the client

Explaining the maintenance process

The client needs to be aware of the system performance monitoring requirements

Motor home with a stand alone solar PV system

HANDOVER

Handover demonstration

Commissioning leaves the system working and ready to hand over to the customer with all the pre-handover checks completed. The paperwork needs to be explained, but more importantly the system needs to be shown to the customer and demonstrated to the new owner. This includes where to gather data on how the system is functioning, and how to turn the system on and off. It is especially important to explain that if the grid goes off their grid connected system will also be shut down until grid power is restored. Simple maintenance checks must also be explained and demonstrated.

Once commissioning is complete the handover process can take place

Handover documentation

The documentation handed over to the customer should be kept in a folder next to the inverter so anyone working on the PV system or other electrical system in future has access to essential information. Some documents are essential for all installations; others depend on the specific installation.

Documentation given to the customer should be kept next to inverter

Essential documentation

Essential documentation includes:

- Electrical installation certificate BS 7671 17th edition with PV amendments and schedule of inspections and test results (or the relevant local or national electrical specification installation, testing and certification requirements)
- Copy of G83/1 SSEG or other relevant installation commissioning confirmation
- System diagram including AC and DC parts
- System operating instructions including procedures to shut down and re-start system
- Inverter datasheet, manual, G83/1 or other relevant test certificate and warranty details
- PV module datasheet, installation manual and warranty
- Installer contact details

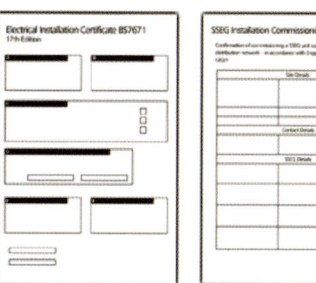 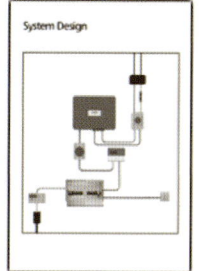 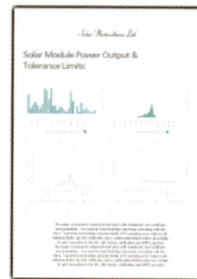

Essential documentation

Optional documentation

Other documents depend on the individual installation designed for the building. If other reports were needed, such as structural engineering, planning permission or listed building consent, they need to be included during handover. Other details can include feed-in tariffs and grant applications. Some systems can be monitored remotely by the customer or equipment manufacturer, and details of how to access the information and use it should also be included.

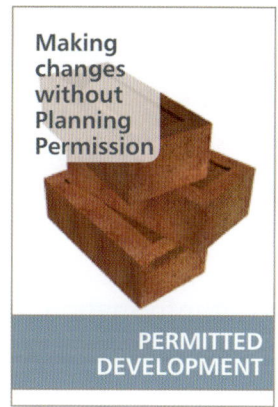 

Additional documentation may be needed depending on the individual installation designed

ACTIVITY 66

What approach to the handover process can considerably improve the whole customer experience and so probably lead to more referrals and more business?

PERFORMANCE MONITORING

Weekly performance monitoring

Regular checks on the performance of the PV system by comparing the expected output with actual output can be satisfying for the customer – but also very useful, as it can identify any loss in performance quickly and therefore help reduce any loss of generation income for grid connected system owners.

Weekly performance monitoring is useful to identify any loss in performance

The inverter can display various items of information, but they can also be monitored remotely. Data can be monitored by the manufacturer as an optional extra, and/or sent directly to the customer's computer. These messages can be public or private as the customer decides. It is also possible to have emails sent if a fault develops.

Inverters can be monitored remotely and display various items of information

Weekly monitoring must always be taken in conjunction with the prevailing weather. Poor performance does not mean the system is faulty! If the weather has been very overcast, performance will be down. Performance may of course exceed expectation if there is a long sunny period too! Part of the documentation may include the expected monthly output from the array, which will give a variable expected output over the year for basic comparison.

Regular checks must always be taken alongside prevailing weather

Annual performance monitoring

The system should be checked every year against the annual figure the installer gave in the documentation, measured in kilowatt hours per kilowatt peak per year. A typical domestic installation could produce, depending on location, about 800 kilowatt hours per kilowatt peak per year. If a kilowatt hour meter is fitted it will give this figure directly, otherwise it can be calculated using the data in the documentation. The figure should be recorded to compare with the next year's performance. If there is a significant difference, even taking into account the weather, it needs to be checked out.

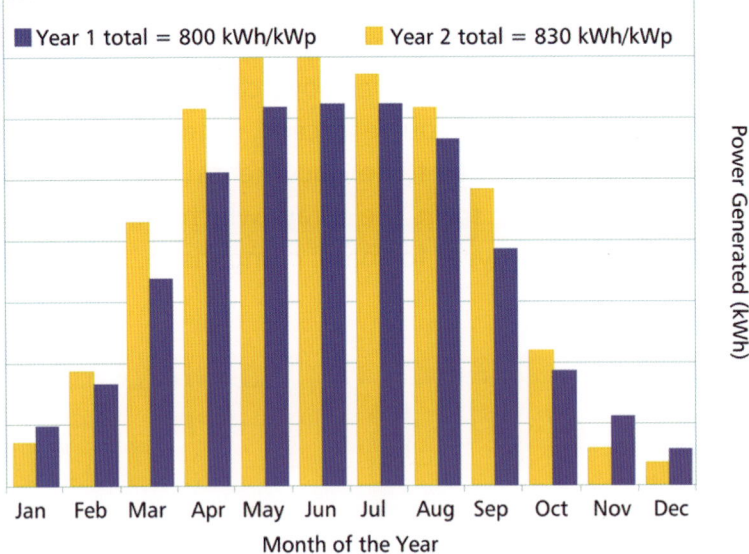

The relationship between power generated and the month of year

ACTIVITY 67

Your customer contacts you and says that their system hasn't performed as well as they were expecting over the last six months. What actions should you take?

Monitoring stand alone systems

Stand alone systems, with batteries, require more monitoring than grid connected systems. The battery voltage needs to be checked often, sometimes twice a day, to ensure they are receiving sufficient charge to maintain electrical power during the night. Batteries may also need topping up with distilled water. Health and Safety regulations apply when maintaining batteries.

Stand alone systems require more monitoring than grid connected systems

ACTIVITY 68

What is the predicted lifetime of a solar deep discharge battery? What actions can you take to maximize this lifespan?

MAINTENANCE

Annual checks

An annual check is advisable, and should be carried out by the installer.

The installer should carry out an annual check

PV array

The PV modules need to be checked for external damage which can affect their performance. They may be shaded by vegetation that has grown over the preceding year, or birds may have built nests which affect the ventilation patterns. Cracks may have developed, or laminates may have degraded, which can be replaced under warranty.

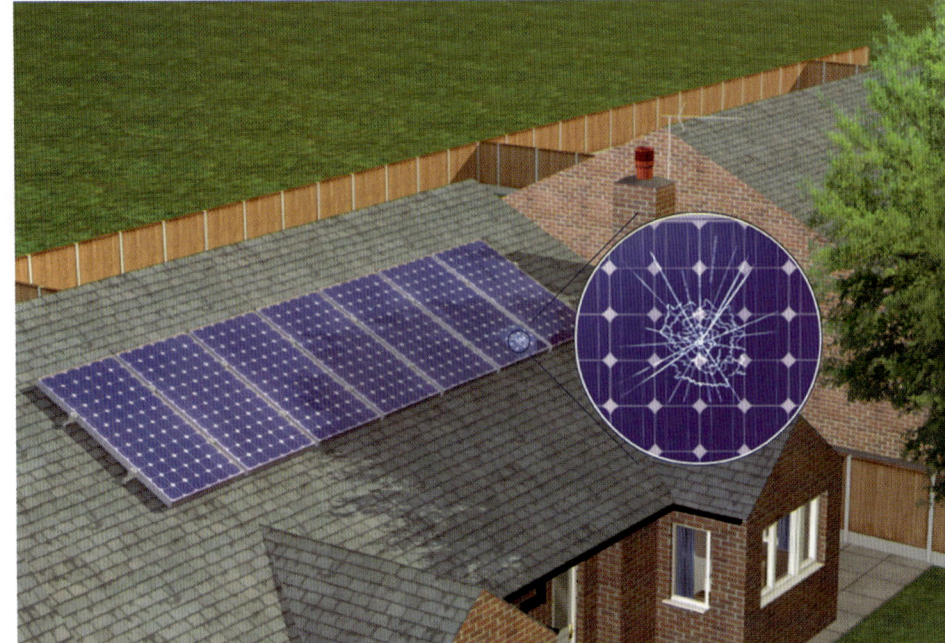

The PV arrays must be checked for external damage

Inverter

The inverter needs to be functioning very efficiently, as it uses some energy converting the DC current to AC current. It needs to have good ventilation and no adverse conditions such as heat and cold or dust interfering with its performance. If it has a fan for hot weather does it run smoothly? The recordings need to be logged and compared with predictions.

The inverter must be functioning very efficiently at all times

General checks

General annual checks on the system include: ensuring the handover documentation is in place by the inverter, the wiring and fittings are OK and all the labels are still in place. If on a feed-in tariff, now is the time to check the readings have been sent to the DNO and payment has been received. Battery systems in particular need to have the condition of the batteries checked – as they do wear out after some years and need replacing.

E-LEARNING

Use the e-learning programme to see more information about annual checks.

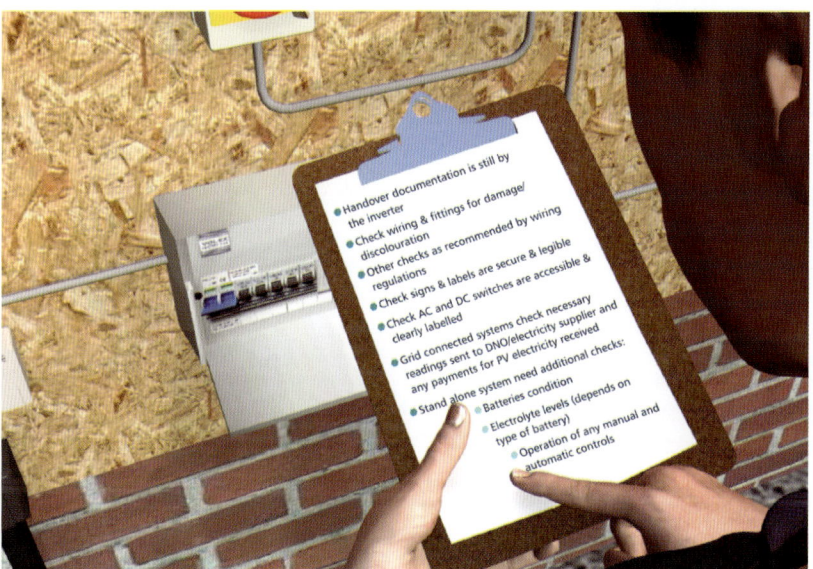

General annual checks

Long term maintenance checks

Long term maintenance can include cleaning the PV array. Normally pitched roofs are self cleaning, but it may be that the pitch is less than usual, or they have built up deposits over the years. When the system was designed cleaning should have been taken into account, as it is a costly and delicate operation. Scaffolding or a cherry picker may be required, so an interval of several years would be normal between cleaning – maybe five years or more. The PV modules are not to be cleaned with high pressure sprays or abrasive substances that may scratch the glass.

Long term maintenance may include cleaning the PV array

ACTIVITY 69

Is maintenance a business opportunity? Can you think of ways of obtaining more income from maintaining solar systems?

CHECK YOUR KNOWLEDGE

1. **Which of the two main types of solar PV installations requires more maintenance?**

    ☐ a. Stand alone PV system

    ☐ b. Grid connected PV system

2. **There are 2 things wrong with the image shown here. Can you name them?**

    ☐ a.

    ☐ b.

3. **It is late summer and the customer records the generation readings every week and notices a gradual but very significantly more than expected decrease in output. There are trees in the garden and the customer's lifestyle has not changed. What could be the cause?**

    ☐ a. Customer forgets the overcast days

    ☐ b. Solar irradiance decreases after midsummer

    ☐ c. PV system has developed a fault

    ☐ d. Trees have grown over the summer

    ☐ e. Customer is using more electricity in the home

4. What is wrong with the image shown here?

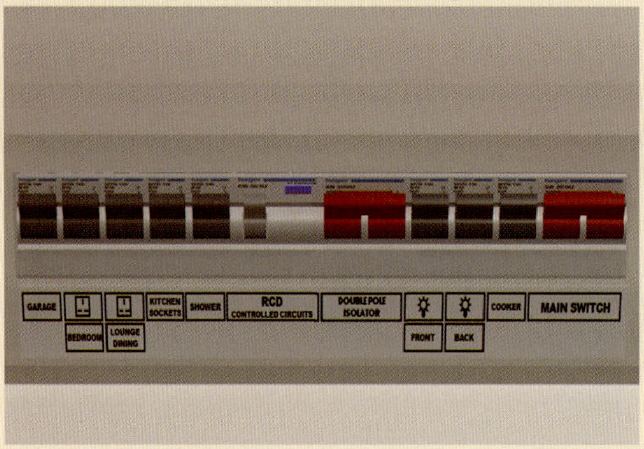

☐ a.

☐ b.

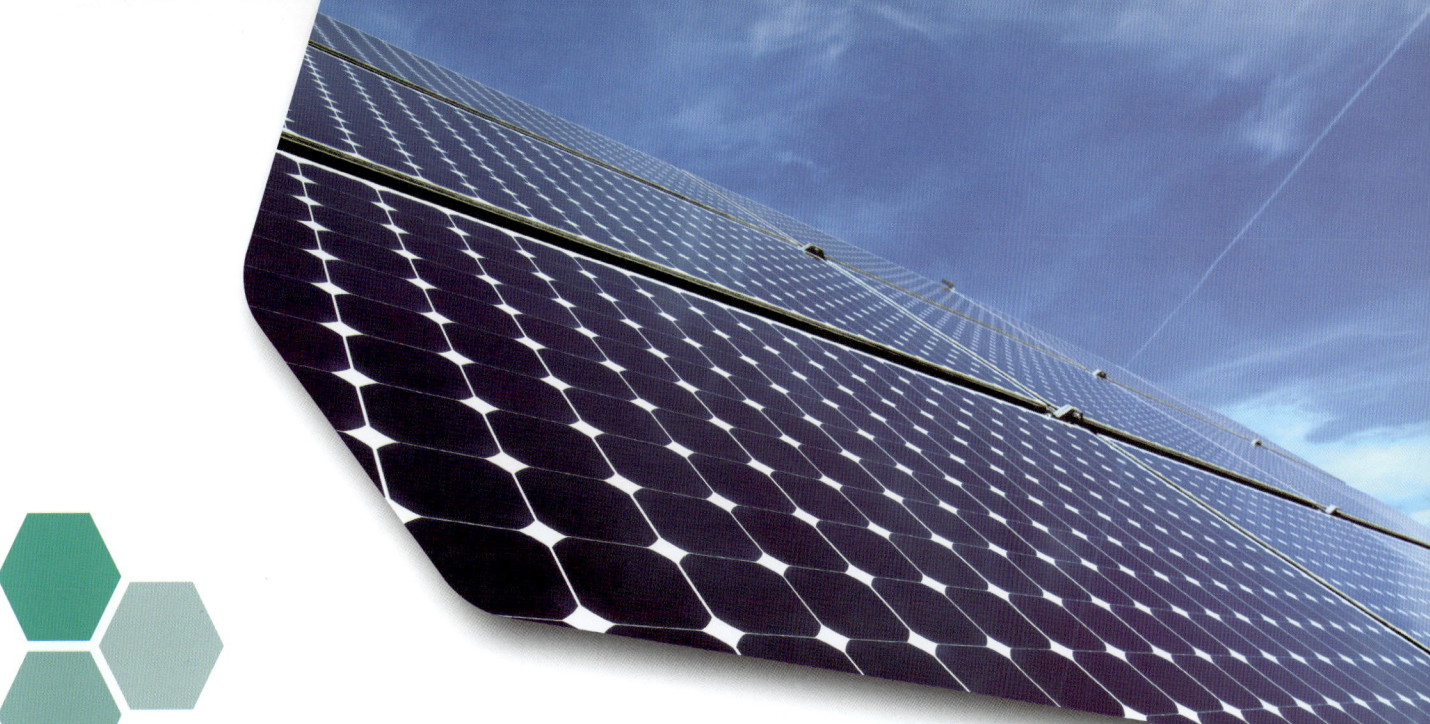

Chapter 12

END TEST

END TEST OBJECTIVES

This end test will check your knowledge of the information held within this workbook.

The Test

CHAPTER 1

1. What type of energy does the Sun supply to Earth?
   - ☐ a. Light
   - ☐ b. Electrical
   - ☐ c. Bio-energy
   - ☐ d. Wind
   - ☐ e. Tidal
   - ☐ f. Water

2. Solar PV installations can produce electricity in excess of domestic requirements during summer on sunny days?
   - ☐ a. True
   - ☐ b. False

CHAPTER 2

3. You are going to be drilling in an extremely dusty environment, and it is likely that a lot of noise will be generated. Which of the following items of PPE should you <u>not</u> use?
   - ☐ a. Safety goggles
   - ☐ b. Safety gloves
   - ☐ c. Earmuffs

4. Which of the following signs indicates the direction you should be walking or driving in?

☐ a.

☐ b.

☐ c.

☐ d.

5. At what angle should a ladder be correctly positioned?
   - ☐ a. 60 degrees
   - ☐ b. 65 degrees
   - ☐ c. 70 degrees
   - ☐ d. 75 degrees
   - ☐ e. 80 degrees

CHAPTER 3

6. What equipment in a solar PV system changes the DC current produced by the PV cells to an AC current used in most domestic circuits?

   ☐ a. Batteries

   ☐ b. Module

   ☐ c. Converter

   ☐ d. Inverter

7. A stand alone system without an inverter has been installed in a holiday home. Which TV can be used in the holiday home?

   ☐ a. TV for AC circuits

   ☐ b. TV for DC circuits

CHAPTER 4

8. What are the characteristics of the following silicon-based PV materials?

   Complete the table by entering the items in the correct column in the table:

   ☐ a. Most expensive

   ☐ b. Cells contain many crystals

   ☐ c. Most efficient

   ☐ d. Cheapest

   ☐ e. Cells made from single crystal

   ☐ f. Most flexible

   ☐ g. Sparkly appearance

   ☐ h. Degrades over time

   ☐ i. Least efficient

Monocrystalline	Polycrystalline	Amorphous

9. Which of the following PV materials are made by a continuous low temperature process during manufacturing?

   ☐ a. Monocrystalline silicon

   ☐ b. Polycrystalline silicon

   ☐ c. Amorphous silicon

   ☐ d. Cadmium telluride

CHAPTER 5

10. What is special about the lead acid batteries for solar PV systems?

   ☐ a. Can be deep cycled

   ☐ b. Can be used in vehicles

   ☐ c. Cannot be isolated

   ☐ d. Produce short bursts of high voltage current

11. By approximately how much can the volume of sunlight on a solar PV installation be improved during summer by using a solar tracking device?

   ☐ a. 10 per cent

   ☐ b. 20 per cent

   ☐ c. 30 per cent

   ☐ d. 40 per cent

   ☐ e. 50 per cent

12. **What information should you know in order to design an efficient solar PV system?**
   - ☐ a. Annual solar radiation for location
   - ☐ b. How many neighbours have solar PV systems
   - ☐ c. Location of nearest power station
   - ☐ d. Measurement of peak demand by household
   - ☐ e. Orientation of roof surfaces
   - ☐ f. Types of trees in locality

13. **How does shading affect the output from a crystalline PV array?**
   - ☐ a. No effect as diffuse daylight is used to maintain output
   - ☐ b. Reduced by the percentage of array that is shaded
   - ☐ c. Reduced to level produced by the shaded cell
   - ☐ d. Reduced to zero as PV arrays cannot tolerate shading

CHAPTER 7

14. **What outcomes do you expect from the pre-installation checks for a grid connected system with on-roof PV modules?**
   - ☐ a. All components available on site
   - ☐ b. Risk assessment for installation work to be carried out
   - ☐ c. Invoice is ready for delivery to customer
   - ☐ d. All components function according to specification
   - ☐ e. Scaffolding in place
   - ☐ f. DNO notified if necessary
   - ☐ g. All faulty items replaced

15. **A grid connected system is being installed. What is the minimum number of DC isolators required?**
   - ☐ a. One DC isolator
   - ☐ b. Two DC isolators
   - ☐ c. Three DC isolators
   - ☐ d. Four DC isolators

CHAPTER 8

16. **When do cables connecting PV modules together have to be connected during installation?**
   - ☐ a. As the framework is built
   - ☐ b. Before modules are installed
   - ☐ c. As modules are installed
   - ☐ d. After modules are installed

17. **Which of the following PV modules can always be positioned at optimum orientation and tilt?**
   - ☐ a. On-roof PV modules
   - ☐ b. Integrated roof modules

☐ c. PV integrated tiles or slates

☐ d. PV modules on flat roofs

☐ e. Pole mounted PV modules

☐ f. Ground mounted PV modules

CHAPTER 9

18. **Where would you find RCD protection in a grid connected PV system?**

☐ a. Generation meter

☐ b. Inverter

☐ c. Isolators (AC and DC)

☐ d. Consumer unit

19. **The charge/load controller for batteries has a DC protection fuse built in.**

☐ a. True

☐ b. False

CHAPTER 10

20. **Which item or items in the list below should be measured for open circuit voltage and short circuit current before commissioning takes place?**

☐ a. PV cells

☐ b. PV modules

☐ c. PV strings

☐ d. PV array

21. **When electrically testing the PV panels during commissioning which of the following are important?**

☐ a. Height above sea level

☐ b. Types of PV panels

☐ c. Weather

☐ d. Temperature

☐ e. Nearest grid connection

CHAPTER 11

22. **Where should the handover documentation be kept?**

☐ a. By the inverter

☐ b. By the generation meter

☐ c. By the distribution board

☐ d. In a safe place at the customer's house

23. **A copy of G83/1 SSEG installation commissioning confirmation is kept with the handover documentation for the customer. Where is the original?**

☐ a. Kept by the installer

☐ b. Sent to the DNO

☐ c. Sent to the local authority environment office

☐ d. Customer also has the original

Answers

The Test

1. A – The Sun's energy is in the form of light energy

2. B – False

3. B – Safety gloves

4. D –

5. D – 75 degrees

6. D – Inverter

7. B – There is no AC current available in the holiday home as there is no inverter

8.

Monocrystalline	Polycrystalline	Amorphous
Most expensive	Sparkly appearance	Least efficient
Most efficient	Cells contain many crystals	Cheapest
Cells made from single crystal		Degrades over time
		Most flexible

9. C – Amorphous silicon

10. A – Batteries need to be deep cycle batteries

11. E – 50 per cent

12. A, D and E

13. C. – Reduced to level produced by the shaded cell

14. A, B, D, F and G

15. C – Three DC Isolators

16. C – As modules are installed

17. D, E and F

18. D – Consumer unit

19. A – True

20. B – PV modules

21. C – Weather and D – Temperature

22. A – By the inverter

23. B – Sent to DNO

Activity Answers

Activity 1

Please note that this is a silly question, as all forms of electricity are potentially dangerous and so full safety considerations must be implemented when dealing with electrical systems. However, it is worth noting that whilst AC tends to throw off the 'live' person, DC holds onto the 'live' person and so can provide a greater electrical shock. As noted, excellent health and safety must always be observed around solar PV and all other renewable energy systems.

Activity 2

The American grid operates on 110 volt 60 Hz AC – whilst the European grid works on 230 volt 50 Hz AC. Therefore, an American or European inverter must produce the form of AC that is suitable for the local grid network.

Activity 3

Energy is used to manufacture and install a PV system. Normally, most of this 'embedded' energy would come from a fossil fuel source and so the PV system would not in the strictest sense ever be zero carbon. However, a well designed and installed PV system will normally payback (produce) the energy used in its manufacture and installation in a fairly short timescale (of two to five years) and so the system over its lifetime will produce far more renewable energy than the fossil fuel used in its production. Therefore, most PV installations are significantly positive in environmental terms.

Activity 4

Similar inclinations and orientations to the UK will be used across northern European countries such as the Netherlands or Denmark. However, further south in countries such as Greece, Italy or Spain where the ratio of direct to diffuse radiation changes to mainly direct rather than diffuse, pointing at the Sun becomes more useful and so angle and orientation becomes more critical and tracking devices that follow the Sun as its location changes in the sky increasingly provide more benefits.

Activity 5

It could produce a maximum nominal 24 volts or 48 amps (at STC). If the cells were wired in two parallel groups of 24 cells, the panel would produce a 12 volt nominal voltage and so in theory be capable of charging a 12 volt battery. However, real world PV panels designed to charge 12 volt batteries typically have a nominal voltage of about 16 volts and have groups of around 36 cells wired together to produce this nominal voltage. If two banks of these cells were wired together in parallel, 72 cells would be required. This is because a 12 volt battery needs a greater than 12 volt potential difference to gain charge.

Activity 6

These are a set of reference photovoltaic device measurement conditions consisting of irradiance of 1 kW/m^2, AM 1.5, and 25°C cell temperature, where AM stands for Air Mass and indicates that the light has travelled through an air distance 1.5 times that found at the equator. By having a laboratory repeatable set of conditions, it makes it possible to compare like for like solar PV panels.

Activity 7

Because of the very significant difference between solar radiation levels in summer and winter in the UK, a system designed to run just on PV would require a lot of energy storage (e.g. batteries) and PV panel 'redundancy'. By redundancy, we mean that there would be many solar PV panels that would be required to power the system in winter that would be sitting there in summer with no function as the overall system is grid disconnected. This could possibly be alleviated by using a joint solar PV and wind system because the wind in the UK blows more in winter and the Sun shines more in summer. This is of course reliant on being in a windy enough location to realize this option. Or the system could be backed up with a biofuel generator set so that when the Sun's energy was low in winter, the system could still be topped up with renewable energy. It would be much easier to design a PV only powered system in Greece, as the solar irradiance levels, whilst they significantly reduce in winter, still produce more solar energy year round to power a PV only system.

Activity 8

A cheap, simple DIY type meter is highly unlikely to be acceptable on a grid connected PV circuit. The householder will probably be receiving a feed-in-tariff or other similar incentive and so accurate metered results are required for these tariff payments. On top of this, the householder and the electrical utility provider will in all likelihood want an accurate reading of electricity generation. Therefore, a high grade meter which is sealed to prevent tampering or interference from third parties will probably be a requirement of either the building regulations or power utility.

Activity 9

As well as grid disconnection in the event of a power cut, it would be a useful feature to grid reconnect on reinstatement of the power from the grid. This would minimize householder interaction with the system. Other useful features of the inverter could be an LCD display showing current and/or historical performance of the solar system and also potentially fault diagnostics. Finally, the inverter might also have Maximum Power Point Tracking (MPPT), which optimizes the power output from the solar array by setting the power production point to the optimal voltage and current.

Activity 10

A transistor chip used in smartphones, PCs and many other modern electronic devices is also normally made from silicon. It differs from silicon PV in that a transistor consists of three layers, p-n-p or n-p-n whilst silicon solar PV is just a double p-n or n-p layer. The double layer used in solar PV is further discussed in this chapter.

Activity 11

Monocrystalline, because of its high efficiency, can be useful when space is at a premium and a certain quantity of power is required in a limited space.

Polycrystalline, with its interesting surface appearance and lower cost, might be useful to appeal to both architects for interesting appearance-based integration possibilities, and to specifiers for its low cost.

Thin film, whilst requiring large surface areas, because of its flexibility, can be employed on a very wide variety of substrates and so can be used in many interesting and innovative situations to supply small amounts of electrical power for low energy demand applications.

Hybrid PV panels are fairly new. They can be tuned to produce power for different light levels and offer some very interesting efficiency opportunities. It will be very interesting to observe future growth in the PV market to see which type of the above technologies is applied where and which technology obtains the most market penetration.

Activity 12

The main disadvantages associated with organic photovoltaic cells are low efficiency, low stability and low strength compared to inorganic photovoltaic cells.

Activity 13

If you search the Internet for home-made solar panels, you will probably initially find more information on solar thermal collectors rather than PV panels. However, there is information to be gleaned. Basically, it's not very easy to set up a silicon processing plant in your own kitchen and the other members of your household probably won't thank you for doing this! However, it is possible to either purchase silicon wafers or acquire some damaged silicon wafers and then find someone who has access to a large laminating machine (somewhat larger in capacity than the A4 laminator in your office) and so laminate up some solar panels.

So whilst this question demonstrates that it is in all probability better to buy professionally made solar panels, if you have gone to the trouble of the Internet-based research, you will now understand the manufacturing processes for PV and solar thermal panels somewhat better than before.

With more direct radiation, it becomes more important to locate the panel in a position that optimizes solar gain. In these more southerly countries, it is often best to use an angle of inclination close to the latitude of the region. In the southern hemisphere, it is important to remember that the panels should now face north rather than south.

Activity 14

There is no fixed answer to this question. The author of this book predicted that amorphous PV panels would develop to such an extent that they would probably dominate the market. This hasn't happened and some commentators say that polycrystalline is now becoming the cheapest form of PV. The author is still optimistic that in the longer term, amorphous PV will prove to be the cheapest/kWh. However, he as well as many others looking to predict the future have been wrong before and he could well be wrong again! And he also notes that cost is not the only driver for reasons to purchase any variety or type of panel over another type of panel.

Activity 15

In the short term, not very much. These racing solar electric cars are custom made dedicated machines that require lots of sunshine to operate and also normally have uncomfortable driving positions so as to improve aerodynamics. However, electric cars are gradually becoming more common and as intermittent renewable energy, including wind and solar power, becomes more common on the national grid, with 'smart' technology, all the batteries in the electric cars could be plugged into the national grid, which could then be used to have a smoothing effect on electricity demand and

supply. Much research is ongoing into these matters. In the dream solution, electric cars would be charged either when the wind was blowing or the Sun shining.

Activity 16

200mm is a requirement to keep the profile of the solar panels relatively low to the roofline. In most houses, the existing roofline acts as a fairly attractive feature and the solar array should integrate into this roofline in an aesthetically pleasing manner. This is why the panels should not protrude more than 200mm above the roofline. It is also why some solar PV suppliers charge a premium for aesthetically pleasing roof integrated PV panels. This is discussed further below.

Activity 17

The roof must be wind, weather and fire-proofed. This indicates that in the event of:

- a shower or vertical driving wind-blown rain
- a burning tree falling on the PV panel mounted on the roof
- a hurricane

the PV panels and their mounting frame maintain the integrity of the roofing structure.

Activity 18

Typically, AC isolators are red and DC isolators are black. However, please note the word typically in this sentence. DC isolators are sometimes red and so there is never a guarantee that either colour means AC or DC, so always inspect the component carefully.

Activity 19

A car battery is sometimes called an SLI battery and SLI stands for Starting, Lights, Ignition. Solar batteries are deep discharge batteries. For lead acid versions, SLI batteries tend to have thin lead plates so that they can provide short blasts of high current for starting an engine whilst solar batteries tend to have thick lead plates so that they can be deeply discharged with minimal long term damage to the battery. SLI batteries will start to deteriorate if they are left to drop below 80 per cent charge levels, whilst solar batteries will only start to deteriorate if the charge levels drop below 20 per cent. However, both batteries will typically only have a life of around three years.

Activity 20

A grid connected inverter just has to turn DC into AC whilst a grid disconnected inverter has to both convert DC into AC and also disconnect the PV system from the grid in the event of a power cut.

Activity 21

First of all, you would have to make sure the multimeters could 'talk' to the laptop, in that they provided an electrical signal that the laptop could register. This could potentially be an analogue or digital signal, although the final recorded data would have to be digital as a laptop is a digital device. If one multimeter was used as a voltmeter and the other as an ammeter and the two results were sampled at a regular basis (every few milliseconds, seconds or minutes depending on the recording timescale and required accuracy), then the two results could be multiplied together to

produce instantaneous power output and with the regular time recording, this power could be converted into energy produced. Perhaps the data could be recorded in a spreadsheet, which then contained formula and graphs to output the energy figures.

Activity 22

A DC cable should say 'Danger solar PV array cable – high voltage DC – live during daylight' every 5m to 10m on a straight run where a clear view is possible between labels.

Activity 23

Resisting wind uplift is normally a much bigger issue than installers first appreciate. For pitched roofs, it is important to make sure that the panels are firmly attached to the rafters (rather than just the battens) and also that the rafters are thick enough to take the load. For example, in many mainland European countries, rafters tend to be thicker and more widely spaced, whilst in the UK, thinner rafters are often employed and the rafters are then closer together. Thinner rafters are more prone to splitting (with fat screws). Therefore, in the UK, NFRC & NHBC have published various documents and guidance on making sure PV panels (and solar collectors) are adequately attached to the roof. Rafters can also be noggined or thickened to offer more support.

Likewise, for flat roofs, it is easy to under estimate the strength of the fixings required. For example, if panels are chemical bolted about 200mm into brickwork below, then during a very strong wind, the panels will act like a wind sock and tend to rip the roof, including the top layer of bricks. Therefore, PV panels are often bolted 600mm to 1000mm into the brickwork. Panels can also be weighted to the roof and again please note that the weight can be in the region of 1 tonne/m^2 panel and many flat roofs are not designed to withstand this weight level.

If in doubt, always consult a structural engineer. It's your business and livelihood that's at risk if you don't.

Activity 24

In-roof solar arrays tend to run hotter and so less efficiently than on-roof mounted arrays because typically they do not obtain as much ventilation cooling as on-roof arrays. This could be overcome by designing lots of ventilation, whilst maintaining the wind, weather and fire-proofing of the installation. This is naturally a challenging, though not insurmountable, design process. However, like most design processes, it can end up as a process of diminishing returns.

Activity 25

1. Due south is in the 70 to 80 per cent region
2. Due south-east is in the 60 to 70 per cent region
3. Due west is in the 50 to 60 per cent region

Activity 26

Most BCC check the voltage of the battery to confirm the state of charge, whether full, discharged or in-between of the battery bank. A good BCC will probably display or indicate current state of charge of the battery. It might do this with an LCD display or LED indicator lights. It will also be efficient, in that it will only consume a small amount of the solar energy to maintain and manage the charge process. A BCC will also protect the solar PV array from damage.

Other useful features could see an adjustable controller so that the on and off charge levels of the system could be set for different battery configurations and also the controller could be adjusted for a variety of different voltages (and/or current ranges). It might be weatherproof or designed to work in other hostile environments. Finally, build quality, long life and reliability are also useful features of a good BCC. Good ergonomics are also important; good design always includes the human interface and makes this simple and effective, which includes good instructions for use (you might have come up with further features not in the notes above and we note that this list is not exhaustive and good design is a creative process where much thought leads to excellent products).

Finally, BCC should be simple to wire with good connectors and the BCC should be electrically safe and secure. Safety is the first consideration in all electrical design.

Activity 27

The main two methods for constructing solar trackers are to use:

- Either a computer programme set up for the date and time of year so that the panel is pointed at the exact position of the Sun in the sky
- Or employ a sensor which measures the point in the sky where the solar irradiance is at its greatest and point the PV panels at this point.

The programme method will always point the array at the Sun. However, pointing at the Sun might not be the highest solar gain; for example, in cases of high diffuse light where lying flat horizontally might be more effective than pointing at the Sun.

The sensor method might cause 'hunting' in that two or more areas of the sky in cloudy conditions might be displaying more diffuse sunlight and so the PV array keeps moving from location to new location and uses energy to move every time, whilst only gaining slightly from each move.

There are methods for overcoming each disadvantage. However, it demonstrates again that solar and renewable energy requires careful planning and consideration on a site-by-site basis.

Activity 28

There is an extensive list of sources of this information. However, by now, we were hoping you might say:

Follow the manufacturer's instructions

And also mention the document 'Photovoltaics in buildings, guide to the installation of PV systems' as two good sources of information to help you design and specify PV systems.

Activity 29

Considering safety first, if there is a power cut, whatever happens your local power circuit must be isolated from the mains grid until it is safe to reconnect back to the grid when the power is fully restored. As discussed earlier, this is to protect any electrical utility workers who are maintaining the grid from accidental electrocution.

Local power can be maintained in two main ways in the event of a mains power cut:

1. through a local battery bank
2. via a mains generator set (genset)

The size of the battery bank will determine its power output and so what can be powered in the event of a power cut. Likewise, the size of the genset will also determine what can be powered. Both a battery bank and a genset will have maintenance, running and replacement costs. The genset will also have noise, fuel, fumes and so nuisance implications. In our observation, most customers only go to the trouble of providing back up in the event of a power cut if either the consequences of a power cut are serious, e.g. a hospital, or an important computer system, or if the power cuts are regular, which tends to only occur in island or politically unstable situations.

Activity 30

In most circumstances when the panels are mounted above a tilt angle of 21°, they will probably have low cleaning maintenance requirements. However for certain considerations, extra care should be taken such as coastal sites (or other sites with plentiful birdlife) where seagulls, droppings might foul the individual PV cells of the array, or in areas of heavy pollution or dust where this might leave deposits on the system. In both of these cases and other situations where dirt could collect on the PV array, cleaning access should be considered. This might be with access systems built into the mounting structure or often provision for cleaning from the floor with pole mounted brushes similar to power washing car cleaning equipment. Or you might even be able to turn this into an income stream for larger arrays which could be cleaned on a contract basis?

Activity 31

The answer is that the system has to be increased in size by 25 per cent. The reason for this is that the system collects 80 per cent of the 100 per cent south-facing system and so 100/80 = 1.25. The system has to be increased in size by a factor of 1.25 which is a percentage increase of 25 per cent. Therefore a 25m^2 system facing directly east would collect the same energy as a 20m^2 system facing directly south.

Activity 32

No, solar radiation is often higher in the tropics than on the equator. This is because equatorial storms mean that cloud cover obscures some of the available solar radiation whilst in deserts such as the Sahara and in Namibia, the solar radiation reaches greater peaks than typically found on the equator. Less than 1 per cent of the Sahara is said to be needed to power Europe. This is because there is over 8000 times more solar radiation reaching the planet than the human race consumes each year and solar radiation is high in the Sahara. This comment is politically sensitive because the Sahara is in Africa and most commentators would suggest that we need to work together across the different continents if we are to provide the world's energy requirements in the future.

Activity 33

Because PV is an electrical system, if one cell on the solar PV panel is obscured, the whole panel's power output significantly drops, whilst if a section of a solar thermal collector is obscured, the performance of the rest of the collector is not affected and so the collector will still produce lots of heat energy.

Please note that this is not an argument for using solar thermal over solar PV systems. Both systems have their own advantages and disadvantages. Basically, if heat is required, then use thermal collectors and if electricity is required, use PV panels. If both heat and electricity are required, make sure that the PV system is mounted where it is less likely to be affected by items such as chimneys, as the solar thermal system is more tolerant of such items.

Activity 34

Solar energy available in any particular location is highly dependent on a number of local variables, especially local hills, mountains and valleys and prevailing winds. Having said this, Britain and Ireland, based as they are in the North Atlantic Drift, have high levels of cloud cover and so more diffuse sunshine. This will be somewhat similar in the western coastal sections of Germany. However, as someone travels east into Germany, continental effects start to influence the climate and this reduces cloud cover throughout the overall year and so slightly increases the proportion of direct to diffuse sunshine (or solar radiation/irradiance if we are to be technically correct). This might increase the overall energy available per m^2 by 100 to 150 kWh per annum for a similar latitude.

Activity 35

It would make the panel behave in a similar fashion to a solar thermal collector in that only the cells that were actually shaded would be lost to solar output. However, the cost implications might not justify the expense of the extra componentry and associated manufacturing cost? As with many innovations, time will tell whether they become established.

Activity 36

Insulation levels are only likely to have a minor rather than major effect on electricity consumption in a gas heating-fired house. However, the electrical running costs of the gas fired heating will be affected by how long the boiler and associated pumps are in operation. This will be very different in an electrically heated house, whether this is via direct electric heating, storage heating or via a heat pump. Insulation will also have a significant effect if the property is partially heated with

electric heating. Heating is used for both space and hot water heating and for the latter cylinder and pipework insulation will save energy consumption.

When the Sun shines it provides lots of thermal energy and so the space heating is less likely to be used. Hot water consumption tends to stay static throughout the year. Therefore, PV energy tends to be generated when space heating is not required. However, energy efficiency tends to lead to energy efficiency. What we mean by this statement is that insulating the property, and making the occupants more sensitive to energy consumption, is said to encourage behaviour change to reduce energy consumption, so all energy efficiency measures should also bring benefits for renewable energy systems. Likewise, fitting solar PV as a renewable energy measure will *probably* tend to change the household's behaviour, making the occupants more sensitive to their energy consumption. Please note the emphasis on the word 'probably' in the last sentence as there is much debate and discussion ongoing in the behaviour change debate and in reality it will affect some householdxs more than others.

Activity 37

In a grid connected system, space availability will normally be the main factor whilst in a grid disconnected system, annual consumption and making sure this requirement is met will be the main factor affecting system size. However, as with all purchases, most systems will normally have to be sized to a budget and so many factors will be taken into consideration.

Activity 38

Efficiency = 100 * Energy out/Energy in

Energy out = 4500/2 = 2250 kWh

Energy in = 20 m^2 * 1000 kWh/m^2 = 20,000 kWh

Efficiency = 2250/20000 = 11.25 per cent

A multicrystalline PV panel has probably been used.

Surface area thin-film panel = 20 m^2 * 11.25 per cent/7.5 per cent = 30 m^2

Addressing the typical household electricity consumption from household to household would indicate very significant differences. For example, young family households are likely to use far more energy than out all day young person households. However, homes across Europe are somewhat similar, with similar levels of electrical appliances. Therefore, besides the effects from average income from richer regions to not-so-rich regions, there are likely to be similar levels of electrical energy consumption across European households (excluding the effects of any electricity used for heating).

Activity 39

Like all engineering components, PV panels have a service life and will also require some maintenance. Previously in this book, there has been a discussion on cleaning or self-cleaning of solar PV. This is one area where some maintenance will

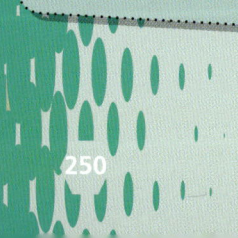

probably occur during the overall lifetime of the panels. However, because PV panels have no moving parts in the traditional engineering sense of the word, and all movement occurs at the sub-atomic level, PV panels are much lower maintenance than most other engineered components. We also note that thin-film panels tend to slowly degrade over time. However this is a slow degrade timescale process and does not require any specific maintenance needs; it just means that the panels produce less energy over time.

Activity 40

This is a vitally important point for the customer because if they do not have faith in the incentive scheme, they will probably not commit to purchase the solar PV system. Therefore, it is important to show the customer that the local, regional or national government has committed in law to maintaining the incentive payments over the lifetime of the scheme. Governments can of course still break the law. However, if they do, the customers have legal redress through the judicial system to hold the government to its legal obligations and the customers are in all likelihoods going to win.

Activity 41

This is an interesting question and presents an interesting moral dilemma. If the government chooses to only pay FiT for grid connected PV, it is saying to households not on the national electricity grid that they have a lower status than connected households. However, paying for grid disconnected households could arguably not reduce the loads and carbon profile of the national grid. Also, grid disconnected households might also be using fossil fuel powered gensets to power their homes and in this case, PV probably makes a big difference to the carbon footprint. In the UK, the national government has addressed this dilemma by choosing to pay the same FiT incentive rate for both grid connected and disconnected households. However all grants and incentive payments can change policy over time so always stay up to date with your local policy framework.

Activity 42

All PV installations should ideally:

1. Look attractive and blend in with the local environment

2. Be electrically safe, secure and sound with correctly sized controls, switchgear, cables and components

3. When mounted on buildings, be wind, weather and fireproof

4. When mounted on the ground, be stable and tamper proof

5. Interface with the local electrical grid or network, whether that is an AC or DC circuit

6. Be low maintenance and reliable

7. Be correctly labelled with relevant information and manuals left on-site

8. Have adequate lightening protection and be properly earthed

9. Have suitable metering arrangements

You might have other features on your list beyond this list above. However, if you addressed the above list with relevance to the particular installation, you will in most circumstances be designing a system that fulfils local requirements.

Activity 43

Slightly different distribution strategies have been employed from country to country. In some countries, it is common to find three phase distribution of power in a local house. In Britain, it is common to find a single phase 230V AC connection in most domestic properties and this connection is often 60A. When the network was installed, it was assumed that each house would draw around 1.5kW maximum at any time. This indicates that if all the houses on a local substation suddenly drew a lot more than 1.5kW at any time, it would probably cause a trip at the local substation.

This scenario is highly unlikely to occur without co-ordinated action on the part of the local inhabitants. PV changes all of this because suddenly, when PV is fitted, the local property becomes a very small 'power station' and now we are feeding electricity in at a local level rather than providing all the supply from a large centralized series of power stations. However, this is causing electrical engineers new challenges as they are having to redesign their electrical networks in the change from large centralized power stations to local distributed power stations. This change will happen over many years and many new developments and changes to the way we distribute, manage and control electrical power are likely to happen as these changes come through.

Activity 44

A generation meter would be fitted between the inverter and the consumer unit in this diagram. The structure of the FiT payments would determine the layout of the different meters. In the UK, a tariff is paid for both generation and export and so in theory, two meters are required to measure both export and generation. In reality, only a generation meter is required as only this is measured and the export tariff is deemed (another word for estimated). In Germany, only the export meter would be required as the FiT is paid just for exported electricity.

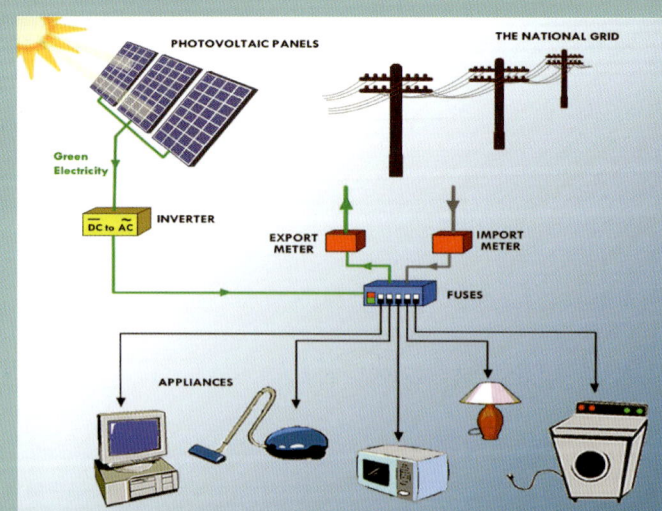

Activity 45

In building mounted PV, the cabling is also mounted on the building. For ground mounted PV, the cabling needs careful consideration and depending on the size and location of the PV system, this will probably involve trenching armoured cable. It is also much easier to track the Sun with ground mounted PV.

Activity 46

We trust this is a simple question to answer as health and safety is often said to be applied common sense. Written risk assessments are useful documents for clarifying thoughts and making sure safety procedures are clear and followed. And of course, you must be concerned about your own and your colleagues, health and safety. You must also take due care and consideration of anyone else who might interface with the PV system, either during installation or also in subsequent use. For example, you would need to make sure that there were no live wires left accessible after installation and that there are no fire risks as just two areas where risk assessment can help.

Activity 47

Customers tend to be reassured by:

- Other customer references and worked examples
- Membership of quality assurance schemes
- Membership of trade associations
- Membership of consumer codes and ethical sales schemes
- Comprehensive insurance cover perhaps including schemes such as insurance backed guarantees
- Membership of any government support schemes
- A high quality website and marketing material
- A company that is professional in every aspect of its approach to business
- Plenty of supporting information so that the end user fully understands their installation

If you implement some or all of the above, your business is more likely to grow and so be more successful.

Activity 48

If the system doesn't function, the whole system has to be disassembled until the fault or faults are located. Therefore, if practical, it is normally better to check individual panels and sections of the system as they are installed.

Activity 49

The free guide, 'Photovoltaics in Buildings, Guide to the installation of PV systems' is useful here. And the particular requirements for cables that should be considered include:

Cables routed behind a PV array must be rated for −15 to 80°C

Cables selected and installed to minimize risk of earth faults and short-circuits

External cables must be UV resistant and waterproof. Much more information is covered in the Guide.

Activity 50

The key here is to allow for air to enter at the base (and perhaps the sides) of the array and pass up the back of the panels and out of the top of the array. This must be done whilst also maintaining the water-proofing layer of the roof structure. Therefore, downward facing ventilation holes at the top of the array that allow the air to escape will assist in this process. And the greater the area of all the ventilation holes, the more effective will be the cooling process as long as the back of the PV array allows the smooth flow of air from the base to the top of the array.

Activity 51

Pitched roof types you might have listed include:

- Double lapped plain tiles
- Single lapped and interlocked tiles
- Slates
- Wood shingles
- Metal standing seams
- Corrugated iron
- Fibre cement roofing
- Plastic sheet roofing
- Thatch roofing

Thatch roof would not be appropriate for a PV installation. Also several other roof types are either brittle or require movement on their seams or other special design features, none of which must be compromised or affected by the subsequent installation of a PV array. If in doubt, check with the manufacturer.

Activity 52

Before fitting any roofing system, or for that matter any system you purchase off a manufacturer or supplier, you should check the small print to make sure what guarantee the supplier is providing, otherwise you can be left with the liability. Therefore, for all roofing systems, check the guaranteed life of the product and also any caveats in the literature. A completely free standing roof frame is far more liable to movement from an unpredictable source and so we would always recommend firmly fixing whatever you are installing so it doesn't easily move. This is not to say that a wind tunnel designed PV array structure won't function as designed. Just remember to always cover all eventualities, and because the roof is exposed to wind, weather and sometimes fire, it is worth being duly diligent here.

Activity 53

No. In the UK, it is very uncommon to reach the $1000W/m^2$ peak capacity of the PV panels and so it is advisable to fit an inverter rated at 80 per cent of peak power. However, the Sun can be much stronger is southern climes and if the local climate reaches a maximum of $900W/m^2$ or $1000W/m^2$, then it would be wise to fit an inverter rated at 90 per cent or 100 per cent of peak capacity respectively.

Activity 54

If there are national grid connection protocols, it means that there are procedures in place to allow PV and other small power systems such as small wind turbines to the national grid. Without these protocols, each individual connection would have to be dealt with on its own merits and this would require much engineering time, knowledge and expertise. In the UK, G83 allows less than 16 amp power generation systems to be fitted without an engineer from the electrical utility being on-site and so simplifies procedures and reduces cost.

Activity 55

In the UK, British Standards is responsible for implementing the standards. DIN covers a similar function in Germany. These bodies often implement European or international standards as well as national standards. Regulatory bodies such as Ofgem often oversee the implementation of these standards, rules and protocols. Certification bodies such as Napit and NICEIC (in England and Wales) allow electricians to self-certify their work to these installation standards. Certifier of Construction schemes cover a similar function in Scotland. As ever, it's vital to work to the local and national electrical framework.

Activity 56

The consumer unit protects the AC section of the electrical circuit in the property. It does not protect the DC section of the PV system. Fuse selection and sizing is discussed further in the guide 'Photovoltaics in Buildings, Guide to the installation of PV systems' and this guide will assist with the selection of string fuses for the DC section of the system.

Activity 57

- Small wind
- Microchip, also called mCHP
- Small hydro power systems
- Anaerobic digestion systems

are the current main four sources of small scale electrical generation.

Activity 58

In such a case advice should be sought from a lightning protection specialist and, where necessary, a lightning protection system installed conforming to BS EN 62305.

Activity 59

An islanding inverter will normally also automatically reconnect to the grid in the event of power reinstatement. Also, an inverter will be electrically insulated so as to make it safe in the case of an electrical fault. Most grid connected inverters will have some form of in-built surge suppression. This in-built surge suppression does not prevent further devices or surge suppression design features from being specified or employed.

When mounting an inverter, it is important to make sure it is firmly mounted onto the supporting wall or structure and also that there is enough access for adequate ventilation to keep the inverter cool. This ventilation requirement should be highlighted to the householder so that they maintain the required space around the inverter.

Activity 60

In the very significant majority of cases, batteries will be used in grid disconnected PV circuits; however, in very low power electrical circuits such as basic solar PV calculators, the electrical power

can be acquired from the ambient light available and so a battery is not required.

Activity 61

BS EN 62305 is called Protection against lightning, it comes in four parts and there are about one million lightning strikes in the UK every decade.

Activity 62

The DC cables should be bundled together so as to avoid the creation of loops in the system. This will act to both shield the cables from inductive surges and, by increasing inductance, attenuate surge transmission. If cables are installed in trunking or conduits, this extra protection must have adequate vents to release condensation or water build-up. DC cable, if not in trunking must have double or reinforced insulation.

Activity 63

This answer is going to look at commissioning the electrical part of the PV system. However, please note that if a fault is located in the roofing or another part of the overall PV installation, other specialist equipment such as roofing etc tooling might be required.

For the electrical commissioning of a PV system:

- For Part P tests, a multifunction tester with a current calibration certificate
- An AC/DC clamp meter with a current calibration certificate
- Pyranometer (solar radiation meter) with a current calibration certificate

are all useful specialist electrical tools. Crimping tools, test probes and leads, labels and test report pads are also all useful tools that will assist in the installation, commissioning and handover process.

Activity 64

Most PV inverters have an LCD display or a data port for connecting to a data acquisition system. Inverters therefore often provide:

- Instantaneous power output
- Long term energy generation

And sometimes, CO_2 saved.

Please note that it is the generation or export meter that forms the final record of the performance of the PV system, and not the PV inverter, so make sure that you have the data your local utility requires recorded from the right place if you are commissioning the system to a protocol such as G83/1 that applies in the UK.

Activity 65

You should also give the customer full operating and maintenance instructions and a full description of the system. Obviously, all handover documentation should be simple, plain and easy-to-follow and should also be accompanied by a system warranty for the system, including component manufacturers and suppliers equipment guarantees. The installation company should act as the reference point for any guarantees and warranty. More information on the handover process is covered in Chapter 11.

Activity 66

Some installers work on the principle of making sure the actual installation is as good as can be and then skimp on the final customer handover. This is a false economy. Working on the principle that the customer is king, the handover process is one of the last moments that the installer has to influence the installation process and so it is worth building time into your quote to allow for a full and comprehensive handover process. Setting aside time for the end user to study the whole system and ask questions they deem appropriate, builds their confidence in the installation and so you and your business. This should lead to less call backs due to lack of understanding and also to more chance of referrals.

Activity 67

First of all, it is important to take the customer seriously and address their concerns (which might be legitimate or not). You need to establish the story behind the call. If you can, get the customer to provide you with the energy data and then compare this to the energy performance in the original quote. Any differences might be down to:

- Dirt build up, leaves, over-shading such as vegetation growth
- Component poor or non-performance such as panels, cables, connectors, isolators or inverters
- Poor design and estimation
- Inaccurate recording by the customer

If it's dirt or leaves on the panels, can the customer clean it themselves? If there is an electrical fault and the system is still under warranty, then it's to your cost. If it's poor design or estimation, then this could have some serious consequences depending on the agreed terms and conditions between you and the customer, so make sure your quotes and 'T&Cs' are good, clear, concise and accurate. And if its inaccurate recording by the customer, look to improve your handover processes so that this doesn't happen again (of course, with some householders, reading meters is a fraught process and in this case, ask them to work off their utility bills which are accurately recorded and produced by a third party).

Activity 68

The battery's lifespan can vary considerably, depending on how it is both used and maintained, including the storage temperature and other factors. If a battery is abused, it is likely to last for less than a year. If a deep discharge battery is only used five to ten times a year for heavy service and is well maintained, it can have a lifetime of over 25 years.

However, this is unlikely to occur in a solar system where the battery will typically be in use daily. For daily use, if well maintained, a solar gel battery should last two to five years and a wet solar battery from four to seven years. To obtain this life, ideally the battery should only be 50 per cent discharged, kept in a well-ventilated room at reasonable ambient temperatures and always topped up (in the case of a wet battery).

Activity 69

Yes, very much so. It is not likely to be the highest turnover part of your business (unless you establish a specialist maintenance business); however, it is useful and regular base income that maintains good relations with your clients and so leads to further upselling opportunities.

As well as regular check-ups, by offering additional services such as panel cleaning or household electrical checks and repairs, maintenance visits can turn into lucrative business opportunities that provide excellent customer service.

Check your knowledge answers

Chapter 1

1. 1. P type silicon
   2. P-N junction
   3. N type silicon
   4. Solar energy
   5. Front metal contact
   6. Free electron
   7. Positive 'hole'
   8. Rear metal contact

2.

Building regulations	**Always**	Sometimes	Never
Planning permission	Always	**Sometimes**	Never

3.

What is the optimum angle for a solar module in the UK?	17°	**35°**	90°
Can solar PV modules use diffuse solar radiation?	**Yes**	No	Sometimes
What is the optimum direction for solar PV modules be facing in the UK?	East	**South**	West

Chapter 2

1. Yes – PPE should still be worn

2. A – You must take responsibility for your safety, however the law requires employers to assess risks, set up control measures and ensure good work practice

3. Plasters, Eye wash, Adhesive dressings, Sterile eye pads, Sterile bandages, Sling, Sterile wound dressings – you should probably add small items of equipment such as safety pins, disposable gloves, tweezers and scissors. However, please note that tablets or medicines should not be part of a first aid kit, as if used improperly they could cause further harm.

4. D, A, E, C, B

5. C – Powder

6. C – There are three faults, the fuse does not match the rating plate, the cable has been repaired, and there are burn marks on the casing

7. C – 50 per cent

8.

Class of Fire	Fire Extinguisher
Class A – wood, paper, textiles, etc	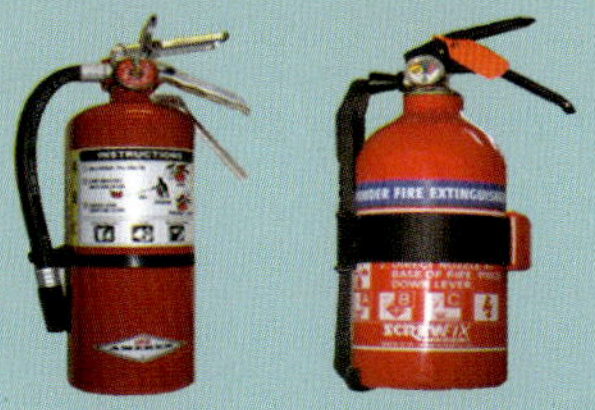
Class B – oil, petrol, paint, etc	
Class C – gas, acetylene, butane, etc	
Class D – metal, magnesium, aluminium, etc	

9.

1.	Try a less risky option
2.	Prevent access to the hazard
3.	Organize work to reduce exposure to the hazard
4.	Issue personal protective equipment
5.	Provide welfare facilities

Chapter 3

1. A. Series
    B. 28V
    C. 28V
    D. 280W

2. A. Parallel
    B. 14V
    C. 20A
    D. 280W

3. A. Inverter

Chapter 4

1. A. Monocrystalline silicon – **15%**
    B. Polycrystalline silicon – **10%**
    C. Amorphous silicon – **5%**
    D. Gallium arsenide – **40%**
    E. CIGS – **10%**

2.

Cell	Image
Monocrystalline cells	

Polycrystalline cells	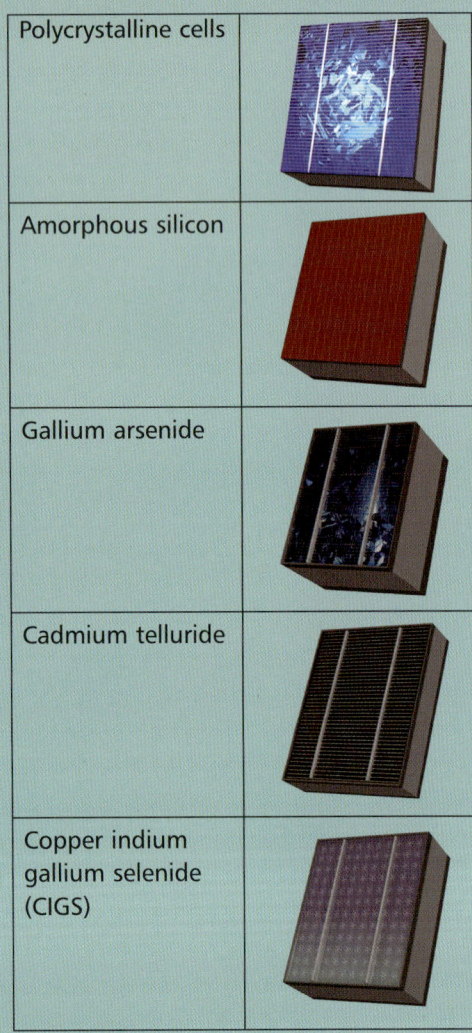
Amorphous silicon	
Gallium arsenide	
Cadmium telluride	
Copper indium gallium selenide (CIGS)	

2.

Mounting System	Image
Solar PV tiles	
Solar PV panels	
Flat roof array	
Ground mounted solar PV panels	
Wall mounted so- lar PV panels	

3. B. Gallium arsenide

Chapter 5

1. 1. DC Appliances
    2. PV Panels
    3. AC Distribution Board
    4. Load and Charge Controller

3. 1. Inverter
    2. Generation Meter
    3. AC Appliances
    4. AC Distribution Board
    5. Import Meter

4. DC cable

Chapter 6

1. 1. B - 2.5% energy loss
   2. A - no energy loss
   3. C - 5% energy loss
   4. F - 20% energy loss

Chapter 7

1.

Combination	Image
Simple one string array	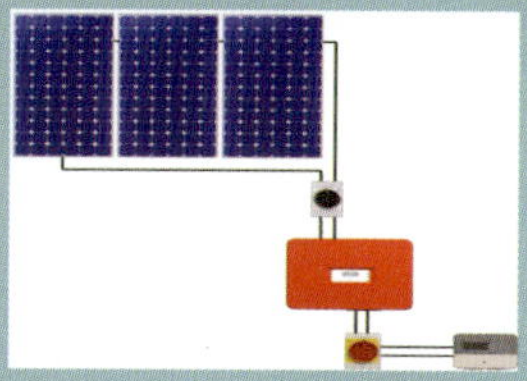
Two String array, one inverter	
Two string array with two inverters	
Three string array with two inverters	

2. D. Their output when joined in an array
3. B. 16 amps

Chapter 8

1. A. PV modules
   B. Inverter
2. C. Wind

Chapter 9

1.

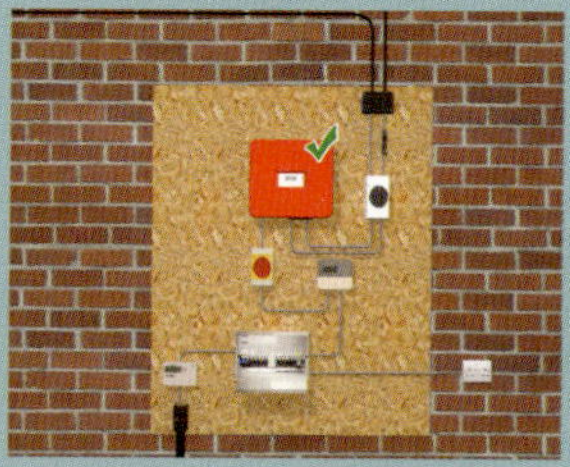

2. A. Overheated cables
   B. PV panels damaged
   C. Fire
   D. Electric shock to anyone working on circuit
   G. Damage to electrical components on the short circuit

Chapter 10

1. B. False
2.

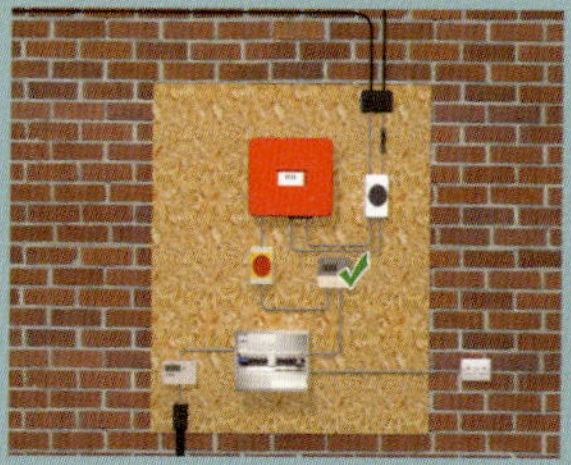

3. D. Qualified electrician

Chapter 11

1. A. Stand alone PV system

2. A. One module is cracked

   B. One module is covered by a leaf

3. C. PV system has developed a fault

   D. Trees have grown over the summer

4. A. Circuit breaker label for the solar circuit is missing

   B. Circuit breaker on the solar circuit is oversized

Glossary

Alternating current (AC) An electricity supply that has an alternating voltage, which changes the current direction many times a second, measured in Hertz (eg 50 Hertz)

Amorphous silicon Silicon where the atoms are not arranged in a crystalline structure. Amorphous silicon can be deposited on glass and other structural materials in a very thin film using less silicon than crystalline technologies

Amp (Ampere, symbol A) The unit to measure the flow of electrons forming the electric current

Array A number of connected PV panels functioning as a single electricity producing unit. The solar panels may be installed on a roof, wall, pole, or ground mounted frame

Autonomous system Alternative name for a stand alone PV system which is not connected to the grid

Back up system An additional source of electricity to ensure an electricity supply when the PV array is not producing sufficient power. Back up can be supplied by the grid, batteries or a generator

Battery Used as a back up system in remote areas as an alternative to the grid or generator. Deep discharge batteries are used in PV systems. Batteries produce DC current

British Standards Institution (BSI) sets quality standards and standard dimensions for equipment and materials. All British Standards start with the letters BS followed by a number

Cable A conductor used to carry current around an installation. Cables are identified by the colour of the installation. DC cabling is also heavy duty and usually Class II

Cadmium telluride A semiconductor used in PV cells, highly toxic in production

Cell A PV cell is the smallest element within a PV module to convert light energy into DC electrical energy

Charge controller A device to limit the charge received by a battery once it is fully charged, and is often combined with a load controller, used to limit the electrical load put on a battery to prevent damage

Conversion efficiency of a PV cell A measure of the efficiency with which the PV cell, panel or array can convert the available solar energy into electricity. Most materials are about 14 per cent, some rare materials have conversion efficiency of 30 per cent

Copper indium gallium selenide (CIGS) A semiconductor used in PV cells

Crystalline silicon A form of silicon in which the molecules are ordered. If the crystals are orientated in one direction, it is known as monocrystalline form. If the crystals are more randomly organized it is a polycrystalline form. Monocrystalline silicon has a better conversion efficiency compared to polycrystalline silicon

Diffuse solar radiation Diffuse solar radiation gives us daylight and can be quite strong. When the Sun's rays hit the Earth's atmosphere some radiation is scattered in all directions creating diffuse radiation

Direct solar radiation Direct radiation is sunshine, which is strongest in the summer at midday because then the angle of the Sun is at its highest

Doping Introduction of other materials or elements that improve the conversion efficiency of silicon

Depletion region An area within the silicon layers of a PV cell where photons of light release sufficient energy to cause electrons to flow, forming the electrical flow

Direct current (DC) Direct current has a continuous voltage and electrons flow in one direction. Solar PV cells and batteries product direct current

DNO Distribution Network Operator. The local area grid operator

Engineering recommendation The recommendations to be followed when installing PV systems connected to the grid. Most relevant are G59/1 and G83/1

Export meter   Meter that records all the electricity generated by the PV array and distributed to the grid

Feed-in tariffs (FiT)   When an electricity supply company is obliged to pay a minimum rate per kWh for renewal energy

Fresnel lens   Use of this lens allows a way of concentrating solar radiation to focus additional light on a PV panel

Gallium arsenide   A semiconductor used in PV cells, relatively more efficient than other substances but also rare and expensive

Generation meter   Meter that records all the electricity generated by the PV array before any has been used by the household, or distributed to the grid

Grid or National Grid   The high voltage electricity network designed to transmit electricity across the country from large power stations to regional distribution networks

Grid connected   A source of renewable energy that is connected to the local distribution network

Guidance notes   There are eight guidance notes to the Briitsh 17th edition IEE wiring regulations published by the Institution of Electrical Contractors

Health and Safety at Work Act (1974) (HASAWA)   All employers are covered by the HASAWA, which places specific duties on both employers and employees to ensure the health and safety of everyone in the workplace.

Heliostat   A mirror that tracks the movement of the Sun to reflect its radiation onto the PV cells/modules

Hybrid cells   Cells made from a mix of crystalline and amorphous material layers tuned to different wavelengths of light

Hybrid system   Use of more than one renewable source of energy for a household, e.g. solar PV and small wind turbine

Inclination   The tilting of something away from a line or surface, or the degree to which it is tilted. For solar systems in the UK, the best angle of tilt or inclination is between 35° to 40° to the horizontal

Information sign   These are green squares with white symbols to give information

Installation checks   Installation checks are made to ensure the installed system first complies with the original design specification for the customer, and then checked for any obvious defects

Integrated applications   Buildings where PV materials form part of the building materials, rather than panels added on a framework

Inverter   A converter that transforms DC voltage and current into AC voltage and current

Irradiance   Instantaneous solar power received on a given surface, measured in W/m^2

Islanding   Situation that occurs if a grid connected PV system continues to supply electricity to the grid during a power cut. This can cause electrocution to technicians working on the grid, but can also create an explosion hazard when power is restored to the grid

Isolator   A disconnect switch that can be used to provide safe isolation of an electric circuit, enabling it to be worked on by an electrician

Joule   The unit used to measure the amount of work done and used in the formula: work done (J) = force (N) × distance. 3 600 000 joules are equal to 1kWh

Kilowatt hour (kW/h)   Unit of energy, usually electrical, equivalent to a device that consumes or generates electrical power at the rate of 1kW for one hour. 1kWh is equal to 3 600 000 joules

Labels   Good practice to label cables and wiring. Especially important to have PV systems labelled to show there are two sources of electrical supply

Load   Anything in an electrical circuit that draws power from that circuit when turned on

Load analysis   A method of calculating household electricity requirements when designing a PV system

Lockable isolator   An isolator that can be locked in the 'off' position with an electrician's own padlock while the system is being worked on

Mandatory sign   These are blue circles with white symbols which tell you what you **must** do

Maximum power point (MPP)   Point on the current-voltage curve where maximum power is produced

Maximum power point tracker (MPPT)   A power conditioning unit that automatically operates the PV generation at optimum MPP under all conditions

Module   A group of solar cells encapsulated in weatherproof glazing, sometimes with a junction box, cables and connectors. Usually factory fitted. Term used interchangeably with 'Panel'

Monocrystalline silicon   Silicon solar cells cut from a single silicon crystal. These are the most efficient form of silicon for PV use

Negative charge Electrons carry a very small negative charge

National Grid, or Grid The high voltage electricity network designed to transmit electricity across the country from large power stations to regional distribution networks

Open circuit voltage (V_{oc}) Voltage across an illuminated PV cell or module when there is no current flowing. It is the maximum possible voltage

Orientation Geographical direction the PV panels are facing. Optimum position is directly South. PV panels can work with other orientations but with loss of efficiency

Panel A group of solar cells encapsulated in weatherproof glazing, sometimes with a junction box, cables and connectors. Usually factory fitted. Term used interchangeably with 'Module'

Parallel connection PV cells connected in parallel will build current

Periodic inspection and testing It is essential to conduct periodic inspection and testing, as the condition of electrical installations deteriorates over time owing to wear and tear, accidental damage and corrosion. The Electricity at Work Regulations 1989 requires that systems are maintained to prevent danger, as far as reasonably practicable

Permit to work A document that specifies the detail of work to be done, when it is going to be done, the hazards involved and the precautions to be taken, and it authorizes the people to be involved in the work

Photon A single bundle of light energy

Polycrystalline silicon Crystalline silicon where the crystals are positioned in a random pattern. Poly-crystalline silicon is slightly less efficient PV material than monocrystalline silicon but much easier to make

PPE Personal Protective Equipment

Prohibition sign These are circular with red crosses through them, and tell you **not** to do something

PV or Photovoltaic The process by which a device converts photons of solar energy directly into electricity

PV Protection Over or under current and voltage protection is required for PV systems

RCD protection Ground fault protection. Full name is a residual current device

Risk Refers to how likely it is that a potential hazard will actually damage your health

Risk assessment Identifying hazards in the workplace then deciding who might be harmed and how

Roof hooks A system for attaching framework to a roof, commonly used with tiled roofs

Roof pitch Angle of the roof slope, which will affect the inclination of any on-roof or in-roof PV system

Semiconductor A substance with conducting properties partway between those of a conductor and an insulator

Series connection PV cells connected in series will build voltage

Short circuit current (I_{SC}) The current flowing freely from an illuminated PV cell or module through an external circuit that has no resistance. The I_{SC} is the maximum current possible

Silicon A common semiconductor used in the majority of PV cells

Smart meters Devices used to measure household or individual item's electrical power consumption

Solar energy The available solar energy is calculated as 1000W/m^2 at the equator at midday

Solar radiation, solar irradiance Solar radiation at a location measured in kWh/m^2/annum. Irradiance is instant solar power in W/m^2. Solar radiation is also sometimes called insolation

SSEG Small Scale Embedded Generator

Stand alone/autonomous system A PV system not connected to the grid

Standards See British Standards

Stand offs A system for attaching framework to a roof, commonly used with slate roofs

Sun path diagrams These are used to calculate available solar radiation and identify any shadow that may affect PV efficiency, or identify alternative sites, during the design stage

Tandem junction PV cells An alternative name for hybrid cells

Tedlar™ Backing material used on many PV panels to reduce weight load

Third generation PV materials New materials developed through nano-technology for increased efficiency and lower cost

Tilt The tilting of something away from a line or surface, or the degree to which it is tilted. For solar PV systems in the UK, the best angle of tilt or inclination is between 35 to 40 degrees to the horizontal

Tracker   A mounting for a small array that pivots the array in one or two directions to follow both the Sun on its east–west path and solar altitude in the sky

Triple junction PV cells   An alternative name for hybrid cells

Voltage   Voltage measures electrical difference between two points. It is the electromotive force that causes electric current to flow in an electric circuit. Symbol V

Ventilation   Ventilation affects efficiency of PV panel, as they are less efficient in higher temperatures

Warning sign   These warning signs are yellow triangles with black symbols – give notice of a particular hazard or danger

Watts   Watts measure electrical power, and when linked to time give units of energy, shown as Watt Hours (Wh). Electrical power is calculated by multiplying current by voltage

Index